T0360309

HSC Year 12 BIOLOGY

REANNA FARLEY | AMI MORROW

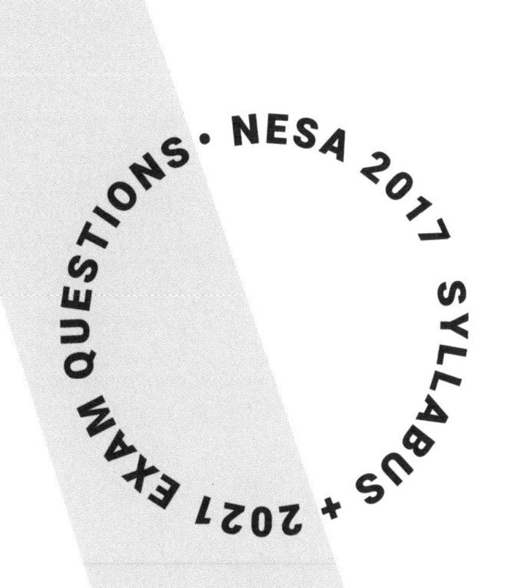

+ 17 topic tests
+ two complete practice exams
+ detailed sample answers

PRACTICE EXAMS

A+ HSC Biology Practice Exams
1st Edition
Reanna Farley
Ami Morrow
ISBN 9780170465250

Publisher: Alice Wilson
Series editor: Catherine Greenwood
Copyeditor: Kay Waters
Series text design: Nikita Bansal
Series cover design: Nikita Bansal
Series designer: Cengage Creative Studio
Artwork: MPS Limited
Production controller: Karen Young
Typeset by: Nikki M Group Pty Ltd
Reviewer: Kirsten Prior

ISBN 978 0 17 046525 0

Cengage Learning Australia
Level 7, 80 Dorcas Street
South Melbourne, Victoria Australia 3205

Cengage Learning New Zealand
Unit 4B Rosedale Office Park
331 Rosedale Road, Albany, North Shore 0632, NZ

For learning solutions, visit **cengage.com.au**

Printed in Singapore by C.O.S. Printers Pte Ltd.
1 2 3 4 5 6 7 26 25 24 23 22

CONTENTS

HOW TO USE THIS BOOK

The *A+ HSC Biology* resources are designed to be used year-round to prepare you for your HSC Biology exam. *A+ HSC Biology Practice Exams* includes 17 topic tests and two practice exams, plus detailed solutions for all questions. This section gives you a brief overview of the features included in this resource.

Topic tests

Each topic test addresses one inquiry question of Modules 5–8 of the syllabus. The tests follow the same sequence as the syllabus, starting with the first inquiry question of Module 5 and ending with the final inquiry question of Module 8. Each topic test includes multiple-choice and short-answer questions.

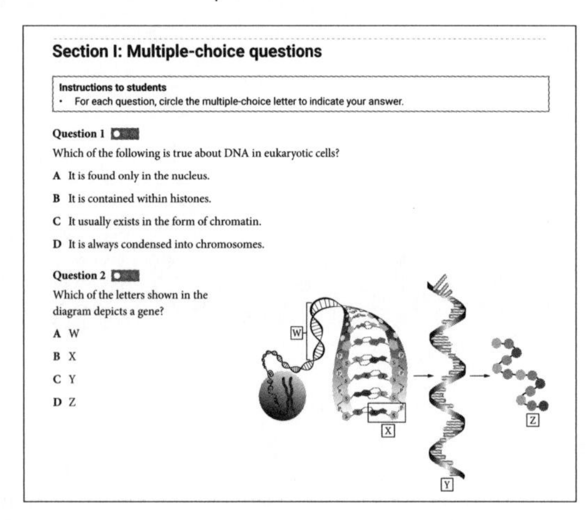

Section I: Multiple-choice questions

Instructions to students
- For each question, circle the multiple-choice letter to indicate your answer.

Question 1

Which of the following is true about DNA in eukaryotic cells?

A It is found only in the nucleus.

B It is contained within histones.

C It usually exists in the form of chromatin.

D It is always condensed into chromosomes.

Question 2

Which of the letters shown in the diagram depicts a gene?

A W

B X

C Y

D Z

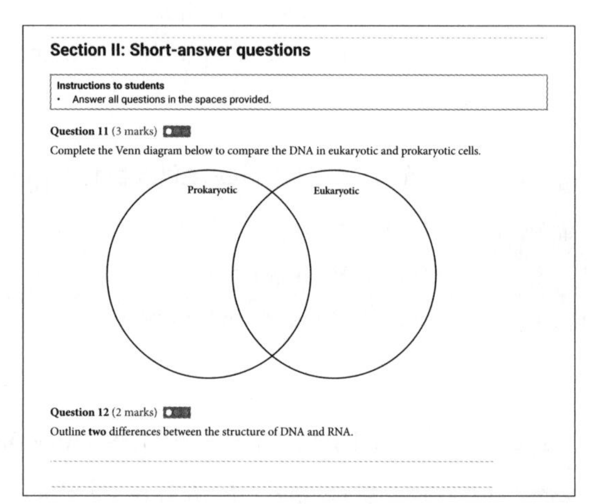

Section II: Short-answer questions

Instructions to students
- Answer all questions in the spaces provided.

Question 11 (3 marks)

Complete the Venn diagram below to compare the DNA in eukaryotic and prokaryotic cells.

Question 12 (2 marks)

Outline **two** differences between the structure of DNA and RNA.

Practice exam section

Both practice exams cover all content from Modules 5–8 of the HSC Biology syllabus. The practice exams have perforated pages so that you can remove them from the book and practise under exam-style conditions.

Solutions

Solutions to topic tests and practice exams are supplied at the back of the book. They have been written to reflect a high-scoring response and include explanations of what makes an effective answer.

Explanations

The solutions section includes explanations of each multiple-choice option, both correct and incorrect. Explanations of written response items explain what a high-scoring response looks like and signpost potential mistakes.

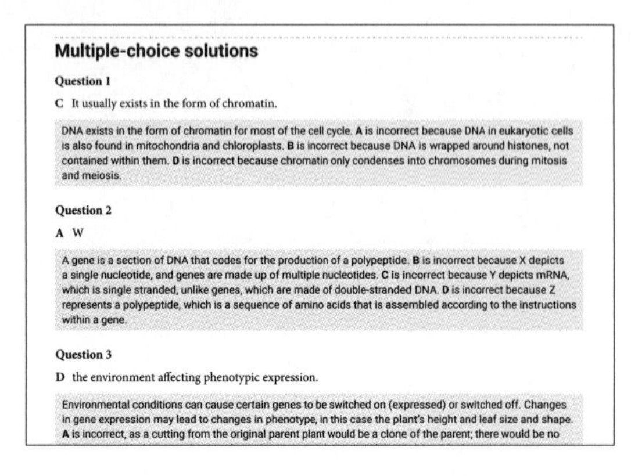

Multiple-choice solutions

Question 1

C It usually exists in the form of chromatin.

DNA exists in the form of chromatin for most of the cell cycle. **A** is incorrect because DNA in eukaryotic cells is also found in mitochondria and chloroplasts. **B** is incorrect because DNA is wrapped around histones, not contained within them. **D** is incorrect because chromatin only condenses into chromosomes during mitosis and meiosis.

Question 2

A W

A gene is a section of DNA that codes for the production of a polypeptide. **B** is incorrect because X depicts a single nucleotide, and genes are made up of multiple nucleotides. **C** is incorrect because Y depicts mRNA, which is single stranded, unlike genes, which are made of double-stranded DNA. **D** is incorrect because Z represents a polypeptide, which is a sequence of amino acids that is assembled according to the instructions within a gene.

Question 3

D the environment affecting phenotypic expression.

Environmental conditions can cause certain genes to be switched on (expressed) or switched off. Changes in gene expression may lead to changes in phenotype, in this case the plant's height and leaf size and shape. **A** is incorrect, as a cutting from the original parent plant would be a clone of the parent; there would be no

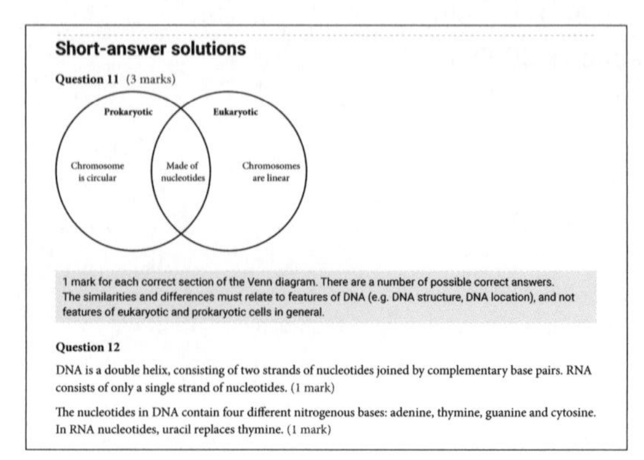

Short-answer solutions

Question 11 (3 marks)

1 mark for each correct section of the Venn diagram. There are a number of possible correct answers. The similarities and differences must relate to features of DNA (e.g. DNA structure, DNA location), and not features of eukaryotic and prokaryotic cells in general.

Question 12

DNA is a double helix, consisting of two strands of nucleotides joined by complementary base pairs. RNA consists of only a single strand of nucleotides. (1 mark)

The nucleotides in DNA contain four different nitrogenous bases: adenine, thymine, guanine and cytosine. In RNA nucleotides, uracil replaces thymine. (1 mark)

Icons

You will notice the following icons in the topic tests and practice exams.

This icon appears with official past NESA questions.

These icons indicate whether the question is easy, medium or hard.

A+ HSC Biology Study Notes

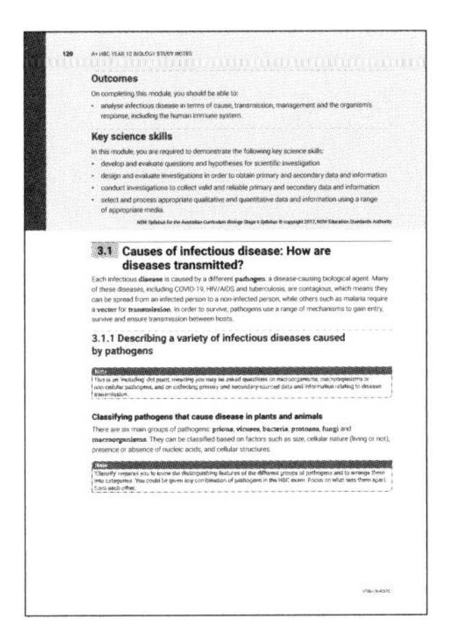

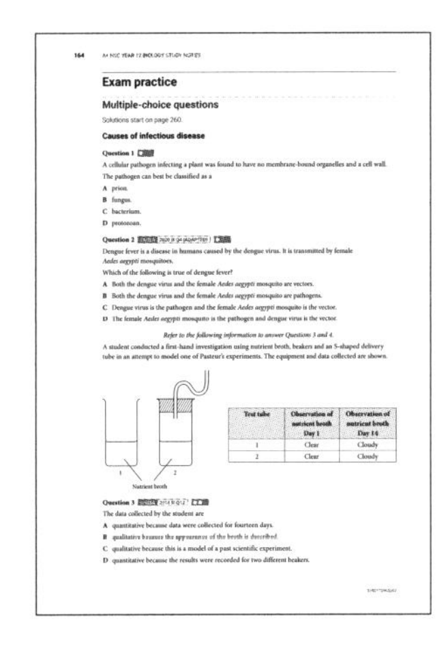

A+ HSC Biology Practice Exams can be used independently or alongside the accompanying resource *A+ HSC Biology Study Notes*. *A+ HSC Biology Study Notes* includes topic summaries and exam practice for all key knowledge in the HSC Biology syllabus that you will be assessed on during the exam, as well as detailed revision and exam preparation advice to help you get ready for the exam.

A+ DIGITAL

Just scan the QR code or type the URL into your browser to access:

- A+ Flashcards: revise key terms and concepts online
- Revision summaries of all concepts from each inquiry question.

Note: You will need to create a free NelsonNet account.

ABOUT THE AUTHORS

Reanna Farley

Reanna Farley is an experienced teacher and marker of HSC Biology. She has taught at Barker College in Sydney since 2004. Reanna has a passion for health and community development and completed a Master's degree in Public Health at the University of Technology, Sydney in 2021. Before this, she spent two years working with a non-government organisation in South Africa, writing public health programs and facilitating health education sessions for underprivileged communities. She also travelled through Uganda with International Medical Relief, supporting health professionals to bring health care to remote communities and providing health education to community members. Reanna enjoys sharing her real-life experience with her students, to deepen their understanding of the syllabus content.

Ami Morrow

Ami Morrow has taught Science and HSC Biology for the past 12 years and is an experienced HSC marker. She is currently teaching Biology at Merewether High School, an academically selective school for gifted and talented students. Ami also works for the New South Wales Department of Education as a Teaching Quality Advisor, where she designs and delivers professional learning to HSC Biology teachers from across the state.

CHAPTER 1
MODULE 5: HEREDITY

Test 1: Reproduction

Section I: 10 marks. Section II: 30 marks. Total marks: 40
Suggested time: 70 minutes

Section I: Multiple-choice questions

Instructions to students

- For each question, circle the multiple-choice letter to indicate your answer.

Question 1

The diagram below shows the process of reproduction in a unicellular organism.

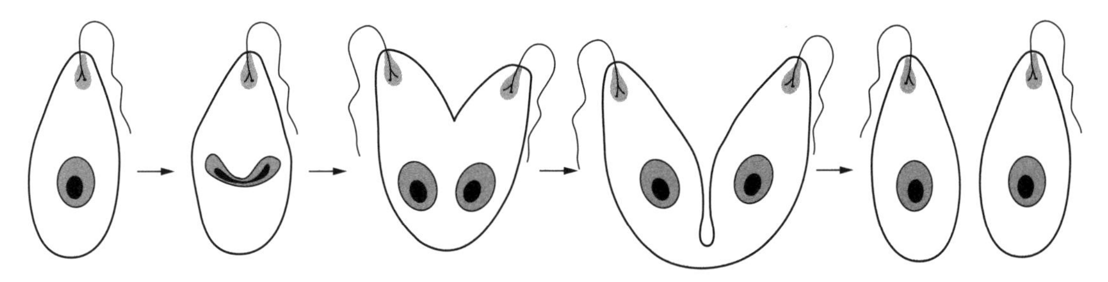

Which method of reproduction is shown in the diagram?

A Spores

B Budding

C Binary fission

D Sexual reproduction

Question 2 ©NESA 2020 SI Q2

Sexual reproduction in plants involves

A pollination caused by dispersal of seeds.

B cloning as it creates copies of the parent plant.

C mitosis leading to the formation of pollen grains.

D fertilisation as a result of fusion of male and female gametes.

9780170465250

Question 3

The diagram below shows a series of processes that occur in human reproduction.

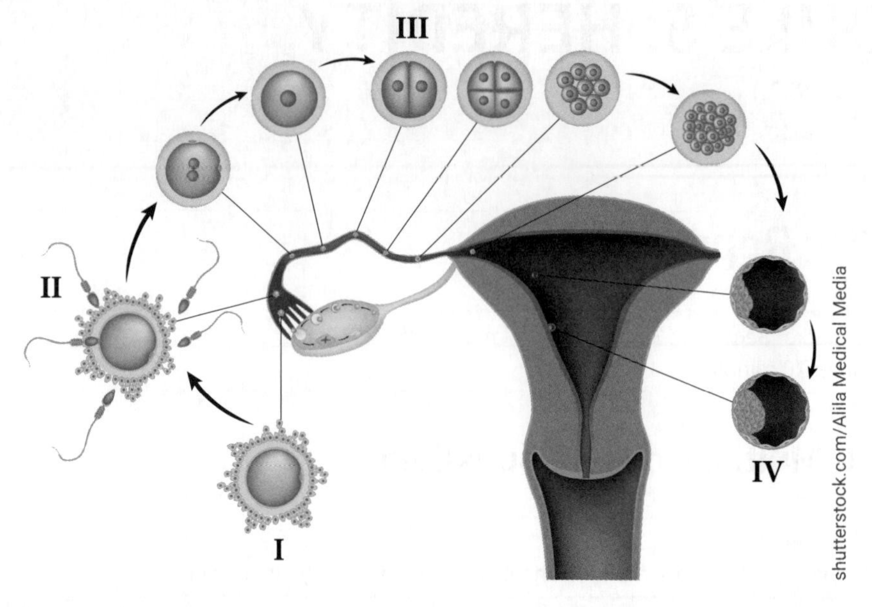

At which point does fertilisation occur?

A I **B** II **C** III **D** IV

Question 4

Sexual reproduction in eukaryotic organisms varies greatly. Which of the following elements do all modes of sexual reproduction have in common?

I Spores II Meiosis

III Gametes IV Implantation

A II and III

B I, II and III

C I, III and IV

D I, II, III and IV

Question 5 ©NESA 2021 SI Q12

The graph shows the levels of three hormones, oestrogen, progesterone and human chorionic gonadotrophin (HCG), measured in the blood of a woman during her pregnancy.

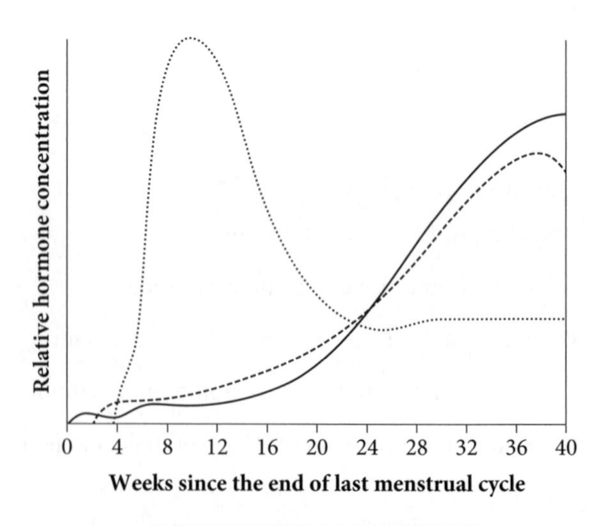

Which statement can be inferred from the graph?

A Birth occurred about week 36.

B Fertilisation occurred at day 0.

C Implantation occurred about week 4.

D The placenta was formed about week 24.

Question 6

Which of the following is always true regarding sexual reproduction?

A It involves copulation.

B It requires two parents.

C It involves fertilisation.

D It produces offspring that are genetically different from one another.

Question 7

In humans, ovulation normally involves the release of a single ovum. In rare cases, several ova may be released at once. If more than one of these ova are fertilised, non-identical (fraternal) twins or triplets may result.

Scientists have recently discovered two genetic variants that appear to be linked to an increased likelihood of having fraternal twins. One variant increases the level of a particular female hormone produced by the brain, and the other increases sensitivity of the ovaries to this same hormone.

The hormone involved is most likely to be

A oestrogen.

B progesterone.

C luteinising hormone.

D follicle-stimulating hormone.

Question 8

The flow chart shows how labour progresses through the action of a positive feedback loop.

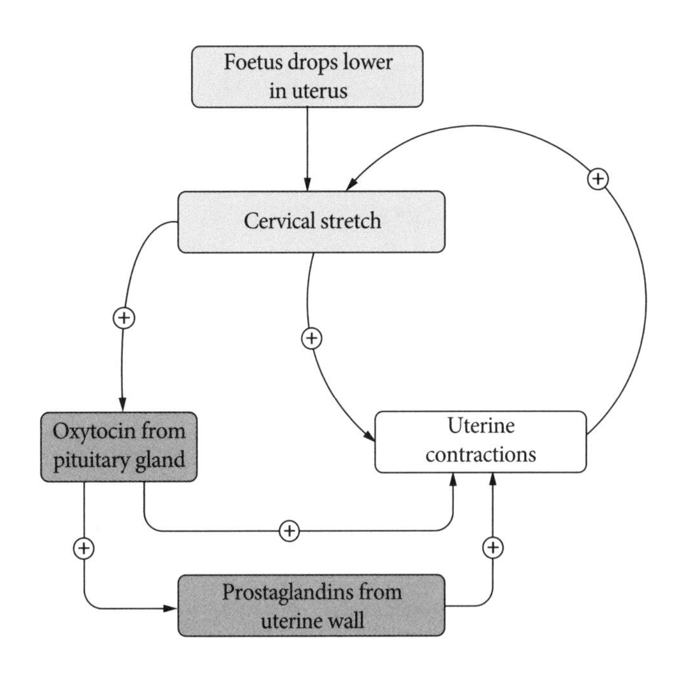

Labour normally begins naturally any time between 37 and 42 weeks of pregnancy. Induction occurs when medical treatment is given to artificially trigger the onset of labour.

Based on the flow chart, which of the following techniques would **not** be used to induce labour?

A A prostaglandin gel is inserted into the vagina, causing the cervix to soften and dilate.

B An injection of an oxytocin receptor antagonist (OTR-A) is given, blocking the action of oxytocin.

C A thin tube called a catheter is inserted into the cervix and inflated with water, applying pressure to the cervix.

D A doctor ruptures the amniotic sac during a vaginal exam, causing the foetus to drop lower into the uterus.

Use the information below to answer Questions 9 and 10.

The diagram below shows the process of alternation of generations in plants. There are two distinct stages: the haploid stage and the diploid stage. In early vascular plants such as ferns, both stages are capable of producing mature, free-living plants.

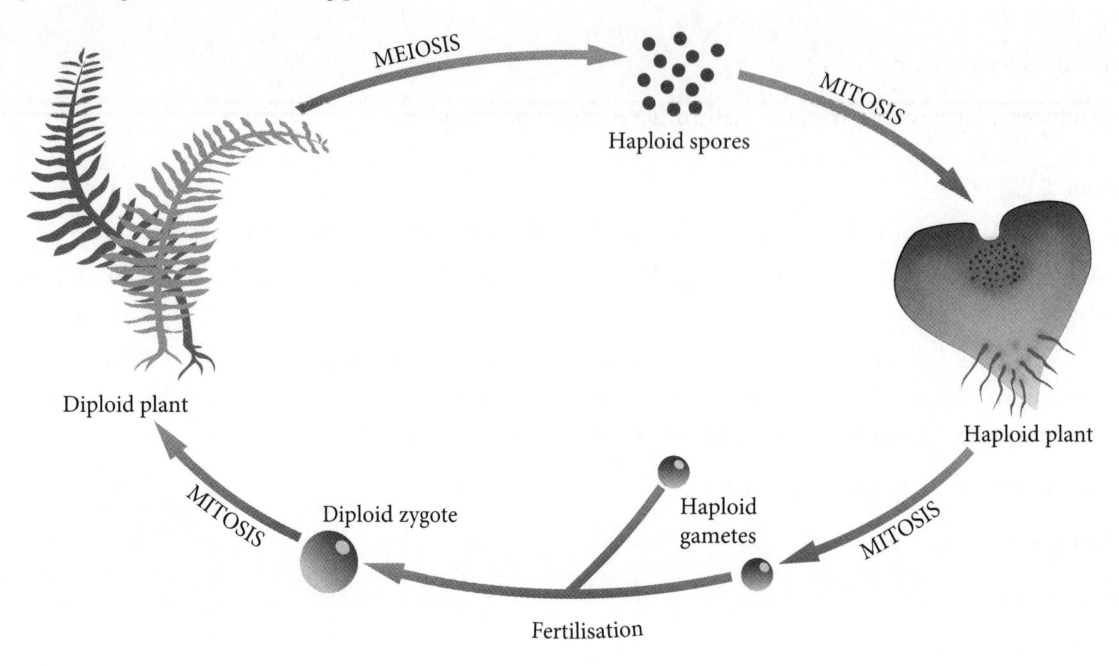

Question 9

Which life cycle stage is found in plants but not in humans?

A Zygote

B Gamete

C Multicellular diploid

D Multicellular haploid

Question 10

As land plants evolved, they moved from spending most of their life cycle in the haploid stage, to spending most of their life cycle in the diploid stage. Modern-day flowering plants live almost completely diploid lives and usually reproduce sexually.

Which of the following is a probable reason for the evolution from haploid-dominant to diploid-dominant life cycles?

A Spores resist drying and cold, so can survive in conditions where seeds may not.

B The diploid generation grows more slowly and takes longer to reach reproductive maturity.

C Spores are light and can travel long distances, allowing for colonisation of new environments.

D In diploid individuals, a second set of chromosomes may mask harmful effects of mutations in the first set.

Section II: Short-answer questions

Instructions to students
- Answer all questions in the spaces provided.

Question 11 (4 marks)

Like many plants, strawberries can reproduce both sexually and asexually. Parent plants reproduce asexually using runners, which are horizontal stems that grow above the soil surface. Each runner produces a tiny plantlet at its end that can be cut away to produce a new plant. Commercially grown strawberries are usually produced this way.

shutterstock.com/Kazakova Maryia

a Explain one benefit to a farmer of growing strawberries in this way. 2 marks

b Occasionally, a farmer may want or need to develop a new strawberry variety. A new variety is created by choosing parent plants from two different varieties and cross-pollinating them. The seeds produced are then planted. The resulting plant is called a hybrid.

The process of developing a hybrid is outlined in the diagram below.

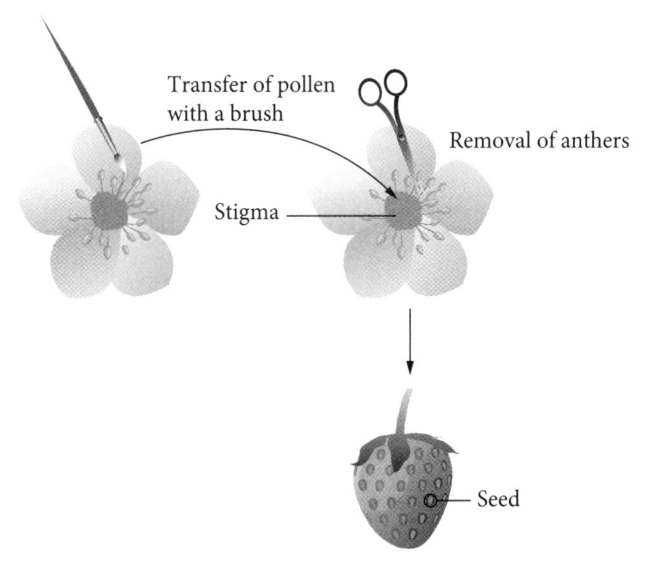

Explain why the anthers are removed from the plant being pollinated. 2 marks

Question 12 (7 marks)

Hydra are simple, freshwater animals from the phylum Cnidaria. They are able to reproduce asexually, as shown in the diagram below.

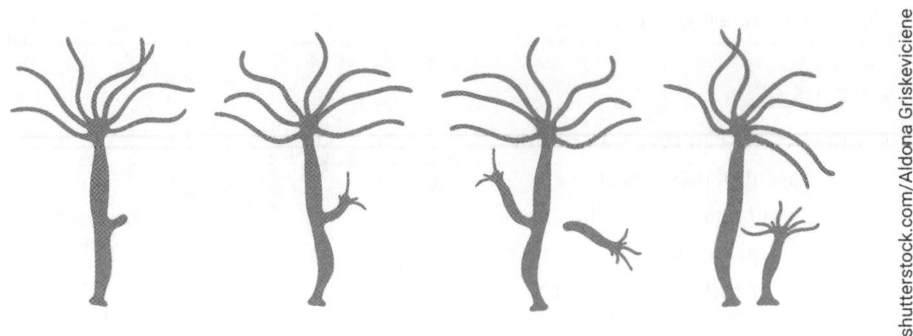

a Name the type of asexual reproduction shown in the diagram above. 1 mark

b Identify another group of organisms that reproduces by this method. 1 mark

c When the environment is favourable and food is plentiful, *Hydra* will reproduce asexually, producing a new offspring every two days.

When conditions are not favourable, they will undergo sexual reproduction. In this process, swellings in the body wall develop into sex organs, which produce gametes. Some *Hydra* species develop only testes or ovaries; gametes of these single-sex *Hydra* will fertilise those of another individual, in a process called cross-fertilisation. Others are hermaphrodites, meaning they develop both testes and ovaries at the same time, allowing the male and female gametes to self-fertilise.

i Place the three methods of reproduction (asexual reproduction, cross-fertilisation and self-fertilisation) in order, based on the amount of variation in the offspring, from highest to lowest. 1 mark

ii Explain how sexual and asexual reproduction in *Hydra* ensure the continuity of the species. 4 marks

Question 13 (5 marks)

The graphs below show the levels of three hormones – luteinising hormone (LH), progesterone (P4) and human chorionic gonadotropin (hCG) – in non-pregnant and pregnant women.

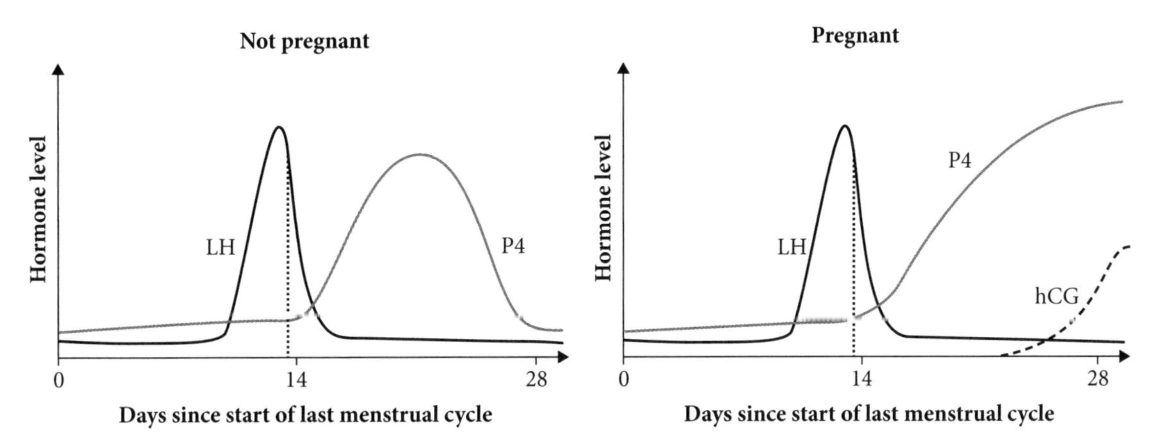

a What key event occurs as a result of the LH surge around day 14? 1 mark

b Account for any differences between the two graphs. 4 marks

9780170465250

Question 14 (7 marks) ©NESA 2020 SII Q25

Students tested the hypothesis that the number of eggs/young produced was greater in animals using external fertilisation than those using internal fertilisation. They obtained the following data from secondary sources.

Mode of fertilisation	Species	Average number of young born or eggs laid in one reproductive cycle	Mean ± SD*
Internal fertilisation	Red kangaroo	1	43 ± 55
	Bush rat	6	
	White tipped reef shark	6	
	Loggerhead turtle	126	
	Red bellied black snake	18	
	Guppy (fish)	100	
External fertilisation	Pouched frog	13	40 ± 32
	Loveridge's frog	20	
	Corroboree frog	25	
	Turtle frog	50	
	Clownfish	100	
	Siamese fighting fish	30	

*SD is standard deviation, which gives a measure of the amount of variation in the data.

a What conclusion can be drawn from the data? Justify your answer. 3 marks

b Justify an improvement to the students' experimental design to test the same hypothesis. 2 marks

c Explain **one** advantage for animals of using external fertilisation. 2 marks

Question 15 (7 marks)

The information below outlines how reproduction in dairy cattle can be manipulated for the purposes of agriculture.

Female cattle do not undergo menstruation. Instead, they undergo a 21-day cycle called an oestrous cycle. The cycle begins at the onset of 'oestrus', a 15- to 18-hour period in which the female is receptive to copulation (mating). During or shortly after oestrus, ovulation takes place. If fertilisation does not occur, the uterine lining is completely reabsorbed by the uterus.

In the dairy industry, cows are selectively bred for their ability to produce large quantities of milk. To maintain lactation, a dairy cow must continually be bred and produce calves. Ideally, a cow should produce a calf every 12 months, with only a 60-day period where they are not producing milk.

Reproduction can be artificially manipulated to allow for optimum efficiency and milk production. Some methods used to manipulate reproduction in dairy cattle are:

- collection of semen from males so that females can be inseminated artificially
- separation and removal of Y-bearing sperm from X-bearing sperm
- devices to detect oestrus
- hormones to induce oestrus and ovulation
- ultrasound or maternal blood tests to detect successful pregnancy in females.

With reference to the information above, evaluate the impact of scientific knowledge on the manipulation of dairy cattle reproduction in agriculture.

Test 2: Cell replication

Section I: 10 marks. Section II: 30 marks. Total marks: 40.
Suggested time: 70 minutes

Section I: Multiple-choice questions

Instructions to students
- For each question, circle the multiple-choice letter to indicate your answer.

Question 1

Which of the following is **not** found in DNA?

A Uracil

B Thymine

C Phosphate

D Deoxyribose sugar

Question 2

The diagram shows a section of DNA.

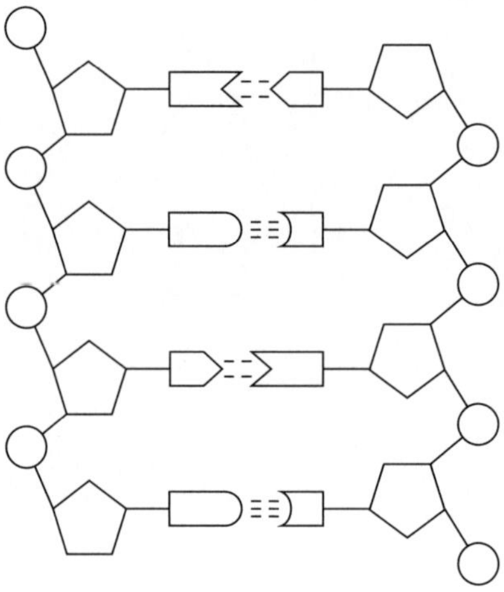

How many nucleotides are shown in the diagram?

A 2

B 4

C 8

D 24

Question 3

Which of the following is true regarding the difference between mitosis and meiosis?

A Mitosis produces only two daughter cells, while meiosis produces four.

B The parent cell in mitosis is a diploid cell, while in meiosis it is a haploid cell.

C Meiosis produces genetically identical cells, while mitosis produces genetically unique cells.

D There are two divisions in the process of mitosis, while there is only one in the process of meiosis.

Question 4

The diagram represents a step in a cell division process.

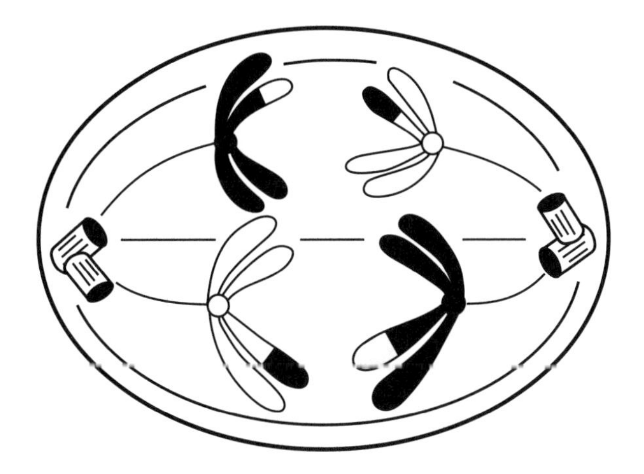

In which cell division process would this step occur?

A Mitosis

B Meiosis

C Cytokinesis

D DNA replication

Question 5

When a cell is preparing to divide, DNA condenses into chromosomes, which can be visualised in a process called karyotyping. Dividing cells are treated with various chemicals in order to stain the chromosomes and fix them in place on a glass slide. The slide is viewed under a microscope and a photograph is taken. The chromosomes are then arranged by size to produce a karyotype, as shown below.

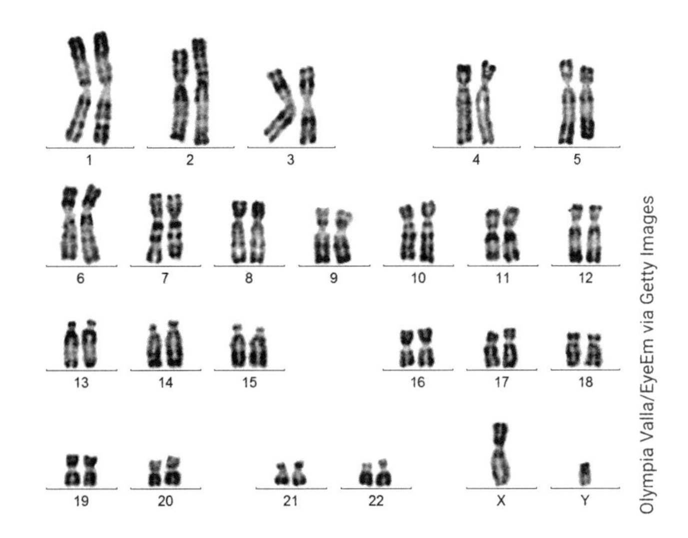

Olympia Valla/EyeEm via Getty Images

The cell used to produce the karyotype shown must be a

A human ovum.

B human sperm.

C somatic cell from a male human.

D somatic cell from a female human.

Question 6

What is shown in the diagram on the right?

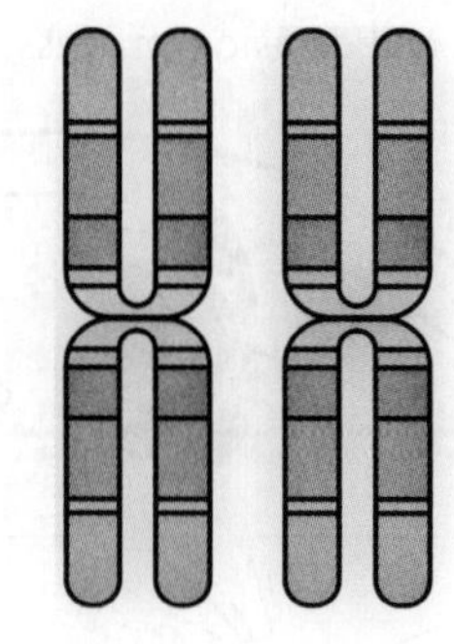

A Four identical sister chromatids

B Four homologous chromosomes

C One pair of identical sister chromatids

D One pair of homologous chromosomes

Question 7

The diagram below shows a model of DNA replication. The original parent DNA strands and the newly synthesised DNA stands are shown.

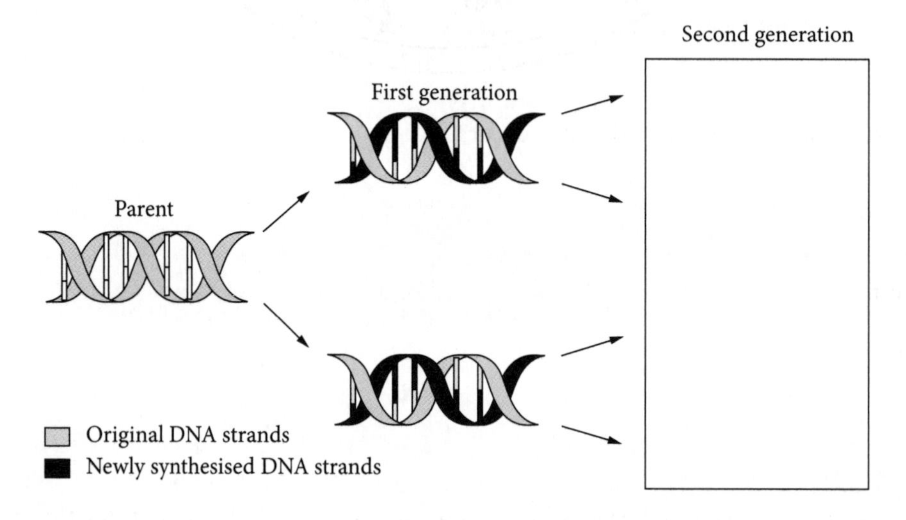

Which of the following shows the correct source of DNA in each of the molecules in the second generation?

A

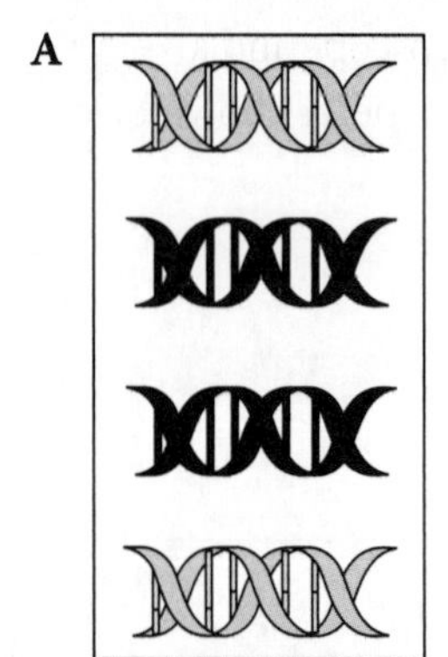

B

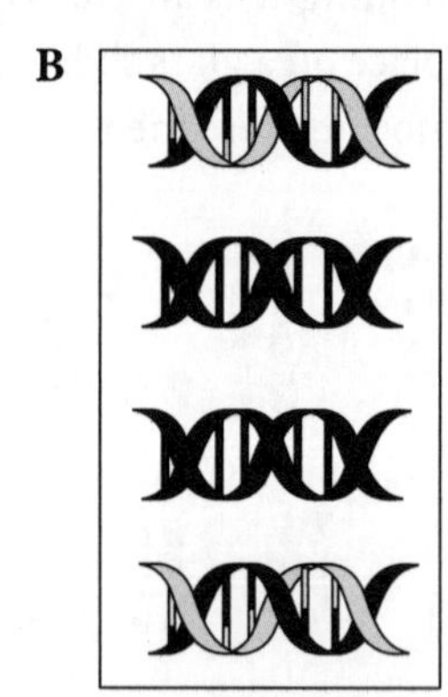

C

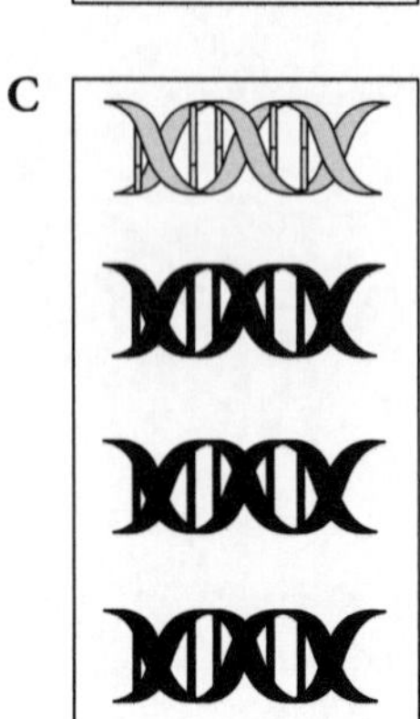

D

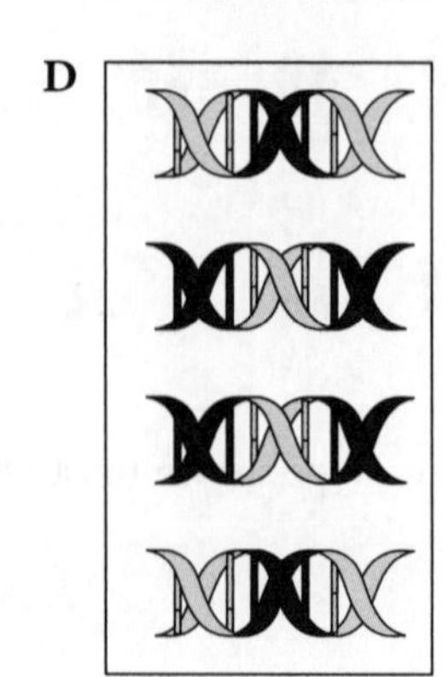

Question 8 ©NESA 2020 SI Q16

Analysis of DNA shows that adenine and guanine always make up 50% of the total amount of nitrogenous bases in DNA.

Which structural feature of DNA does this provide evidence for?

A DNA is helical in structure.

B DNA is always a double-stranded molecule.

C DNA always has adenine paired with guanine.

D DNA is made up of equal amounts of nitrogenous bases.

Use the information below to answer Questions 9 and 10.

The cell cycle is the series of events that take place within a cell as it grows and divides by mitosis. The cell cycle consists of distinct phases: G1, S, G2 and M. The G0 phase is a resting state; cells in this phase are not actively dividing but are performing their assigned role within the body.

The diagram shows the phases of the cell cycle and the genetic contents of the cell at each phase.

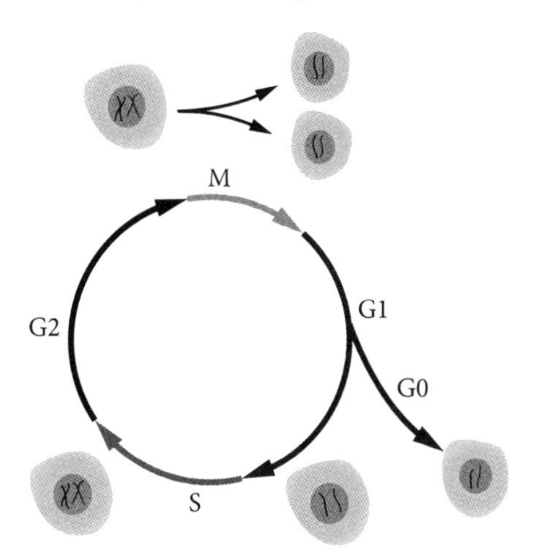

Question 9

During which phase of the cell cycle must DNA replication occur?

A G1

B S

C G2

D M

Question 10

Why is the cell cycle essential for the continuity of multicellular species?

A It is required for sexual reproduction.

B It allows prokaryotes to reproduce asexually.

C It allows organisms to grow and survive to reach reproductive maturity.

D It introduces new characteristics into offspring as a result of recombination.

Section II: Short-answer questions

Instructions to students

- Answer all questions in the spaces provided.

Question 11 (2 marks)

The diagram below shows a model of DNA structure.

Label the following components on the diagram: sugar, phosphate, base, hydrogen bond.

Question 12 (6 marks)

The diagram below shows a process that occurs in eukaryotic cells.

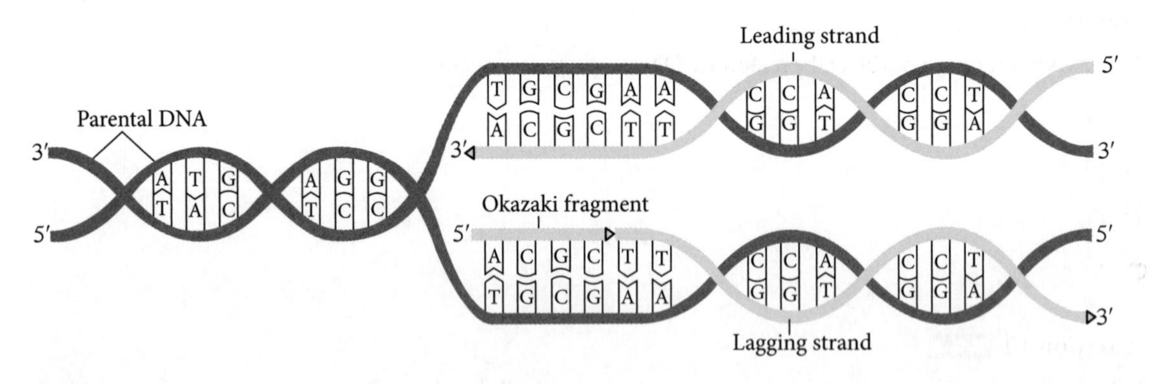

a Identify the process shown in the diagram. 1 mark

b Describe the process shown in the diagram using a series of numbered steps. Ensure that you include the enzymes involved. 5 marks

Question 13 (5 marks)

The images below show plant cells in various stages of mitosis.

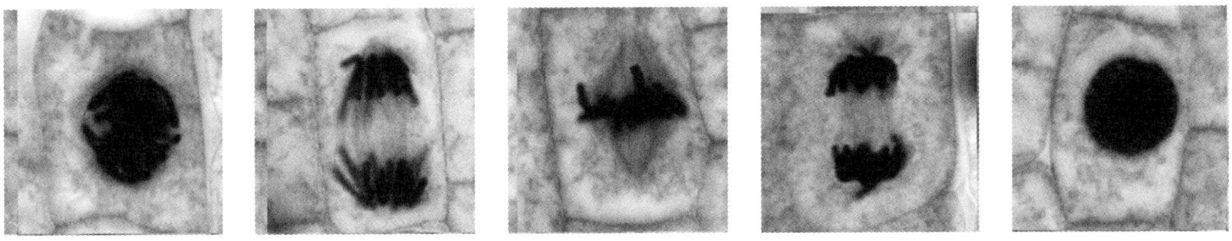

A B C D E

Steve Gschmeissner / Science Photo Library

a List the images in the order that they would appear during the process of mitosis. 1 mark

b Choose **two** stages and outline what can be seen in each stage. 2 marks

Stage:

Stage:

c Explain why DNA must be replicated prior to mitosis. 2 marks

Question 14 (5 marks)

The diagram below shows a cell with a diploid number of 4 undergoing meiosis.

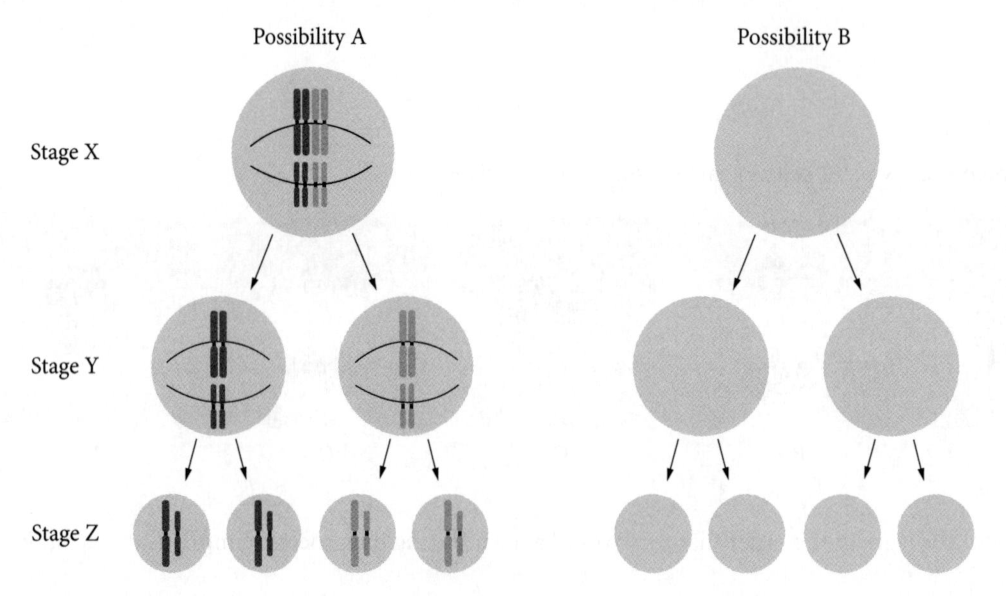

a Besides DNA replication, identify **two** events that would have occurred prior to stage X. 2 marks

b Complete Possibility B on the diagram to show how independent assortment could result in another chromosome combination in the gametes. 3 marks

Question 15 (5 marks) ©NESA 2017 SII Q24 (ADAPTED)

a Three genes are arranged along a homologous pair of chromosomes as shown.

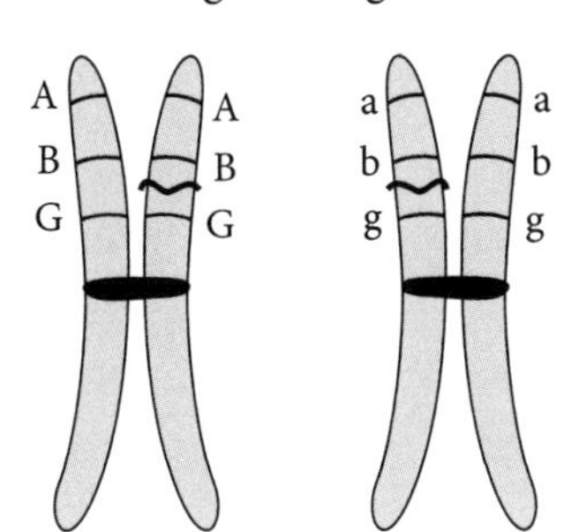

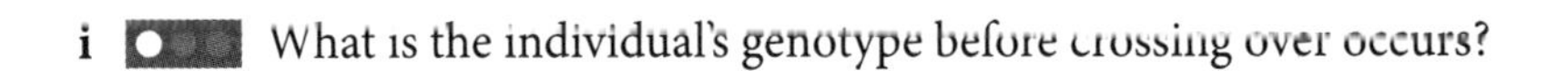

i What is the individual's genotype before crossing over occurs? 1 mark

____

ii Label, on the diagram below, the alleles after crossing over has occurred. 1 mark

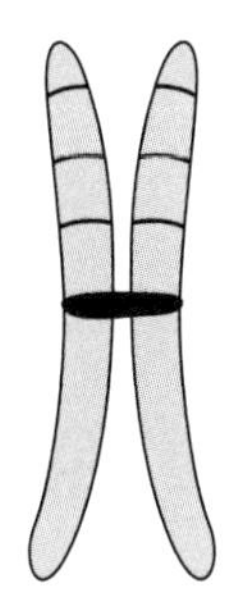

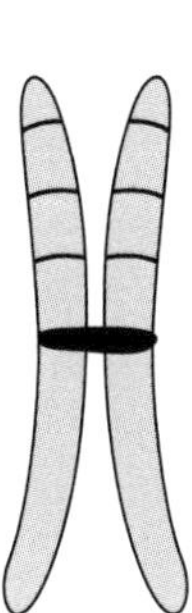

b Explain the effect of crossing over on the amount of variation in the offspring. 3 marks

____

____

____

____

____

____

Question 16 (7 marks)

Explain how independent assortment and random segregation contribute to the continuity of species.

Test 3: DNA and polypeptide synthesis

Section I: 10 marks. Section II: 28 marks. Total marks: 38.
Suggested time: 70 minutes

Section I: Multiple-choice questions

Instructions to students
- For each question, circle the multiple-choice letter to indicate your answer.

Question 1

Which of the following is true about DNA in eukaryotic cells?

A It is found only in the nucleus.

B It is contained within histones.

C It usually exists in the form of chromatin.

D It is always condensed into chromosomes.

Question 2

Which of the letters shown in the diagram depicts a gene?

A W

B X

C Y

D Z

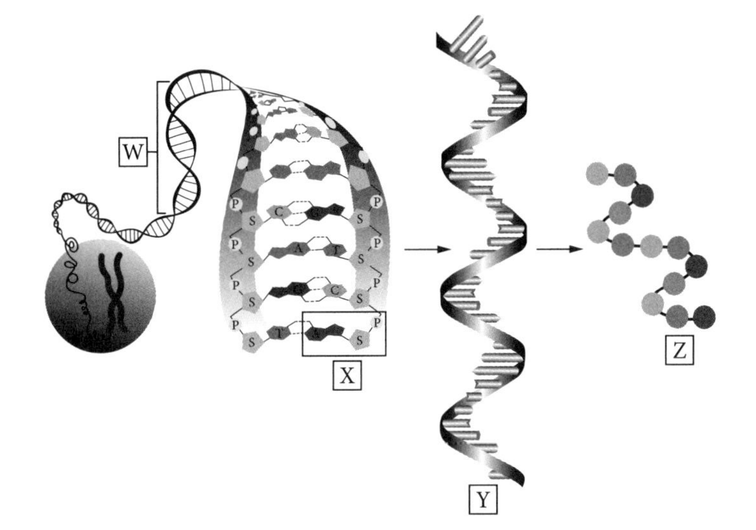

Question 3

A plant was found growing at high altitude in the alpine region of New South Wales. It was stocky in appearance, with small, thick leaves.

A cutting of this plant was taken and grown in a greenhouse, in ideal growing conditions. The cutting grew much taller than the original plant, and its leaves were larger and thinner.

This change in leaf size is an example of

A genetic variation within a species.

B the effect of mutations on phenotype.

C evolution to better suit the environment.

D the environment affecting phenotypic expression.

Question 4

The primary structure of a protein represents the

A linear sequence of amino acids.

B active, functional form of the protein.

C three-dimensional structure of the protein.

D sequence of bases in the mRNA used to produce it.

Question 5 ©NESA 2021 SI Q9

Streptomycin is an antibiotic that kills bacteria by interfering with the function of their ribosomes. The primary effect of the antibiotic is that it prevents the bacteria from producing

A tRNA.

B mRNA.

C amino acids.

D polypeptides.

Question 6

Human serum albumin is the most abundant protein in blood plasma. It consists of a single polypeptide chain containing 585 amino acids. Based on this information, what is the highest level of protein structure possible in human serum albumin?

A Primary

B Secondary

C Tertiary

D Quaternary

Question 7

Viruses cannot carry out polypeptide synthesis. Instead, they must produce viral proteins by injecting their genetic material (either DNA or RNA) into host cells. Once inside, the genetic material is integrated into the genome of the host cell and is converted by the process of polypeptide synthesis into viral proteins.

Which of the following would explain why viruses cannot carry out polypeptide synthesis?

A They lack genes.

B They lack ribosomes.

C Their genetic material is not made of nucleotides.

D They lack a nucleus to house their genetic material.

Question 8

A triplet of bases in the coding sequence of DNA is GTA. The anticodon on the corresponding tRNA would be

A CAT.

B CAU.

C GTA.

D GUA.

Question 9 ©NESA 2020 SI Q20 (ADAPTED)

This chart illustrates three correlation patterns indicating the influence of genes and environment on different traits in individuals.

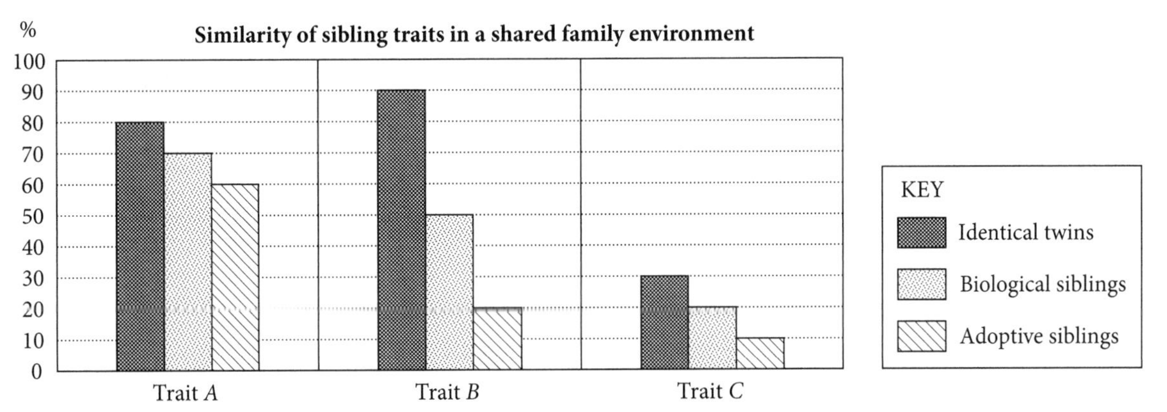

http://wikipremed.com/mcat_course_psychology.php?module=1§ion=16

Based on the data, which of the following statements is correct?

A Identical twins are always genetically identical.

B Genes have a greater effect on Trait *A* than Trait *B*.

C For Trait *C*, both genes and the family environment have a large effect on phenotype.

D For Trait *B*, genes have a greater effect on phenotype than the family environment does.

Question 10

Genes contain regions of DNA, called introns, that do not code for polypeptides. After transcription, introns are removed and the remaining sections, called exons, are joined together. This process is called splicing. In some cases, certain exons may also be removed during splicing, as shown in the diagram below.

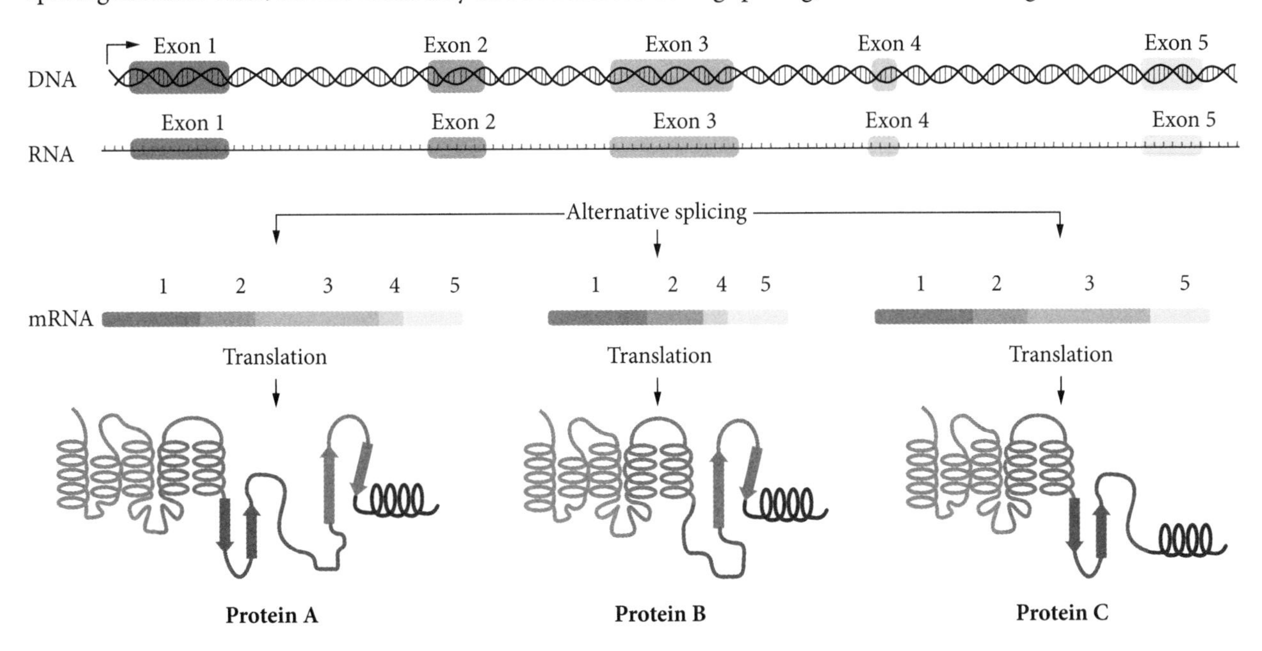

Which of the following is true regarding splicing?

A It removes introns from the DNA.

B It removes all amino acid coding sections from the mRNA.

C It allows for many different polypeptides to be produced from a single gene.

D It ensures that all cells expressing a particular gene produce the same protein.

Section II: Short-answer questions

Instructions to students

- Answer all questions in the spaces provided.

Question 11 (3 marks) 

Complete the Venn diagram below to compare the DNA in eukaryotic and prokaryotic cells.

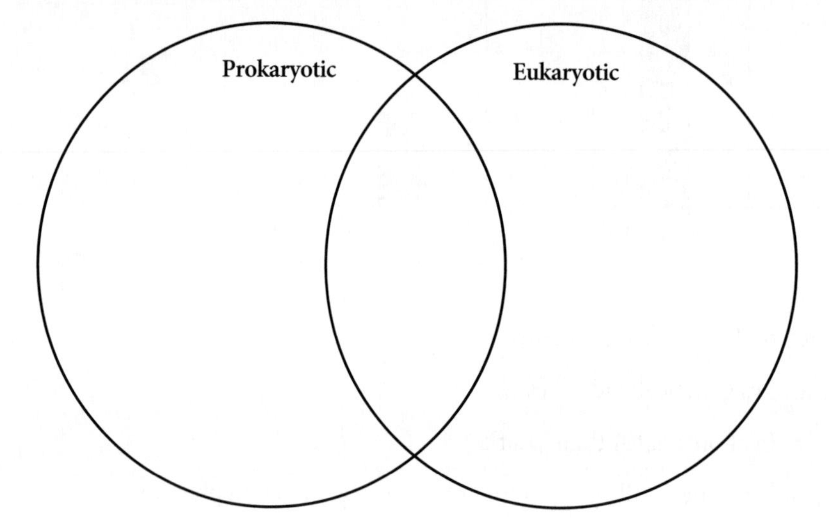

Question 12 (2 marks)

Outline **two** differences between the structure of DNA and RNA.

Question 13 (3 marks)

Describe the relationship between amino acids, polypeptides and proteins.

Question 14 (7 marks)

The diagram below shows the process of transcription.

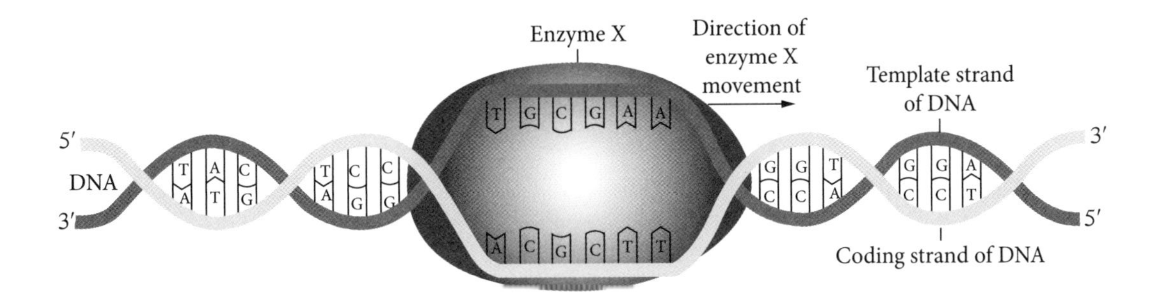

a Identify enzyme X. 1 mark

b On the diagram above, draw the mRNA that is being produced by enzyme X. 3 marks

c Compare the base sequence in the mRNA to that of the coding DNA strand and the template DNA strand. 3 marks

Question 15 (6 marks)

The diagram below shows part of the process of translation.

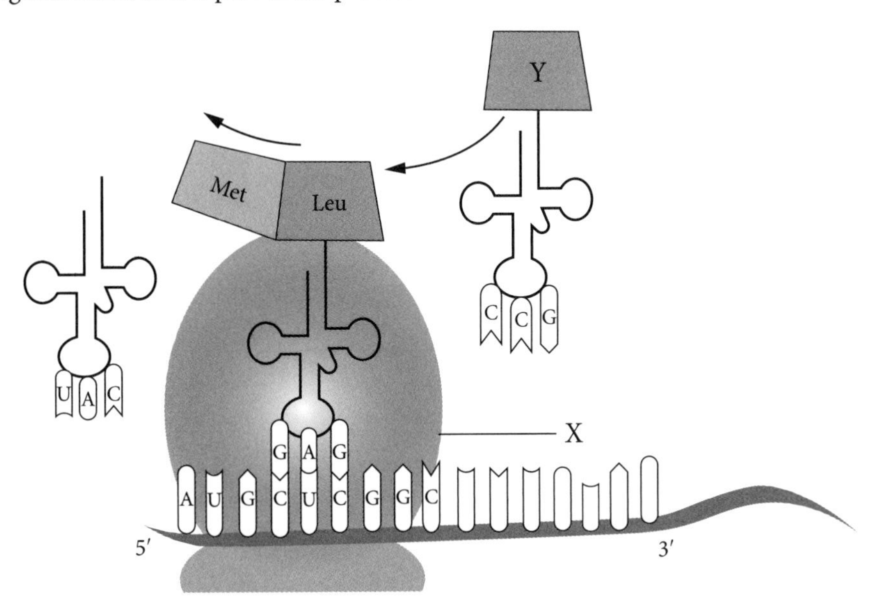

a Identify structure X. 1 mark

MODULE 5 – TEST 3

b Describe how the structures of mRNA and tRNA allow for the process of translation. 4 marks

____

____

____

____

____

____

____

____

c Use the mRNA codon table below to identify structure Y. 1 mark

____

Second base

First base	U	C	A	G	Third base
U	UUU UUC } Phe UUA UUG } Leu	UCU UCC UCA UCG } Ser	UAU UAC } Tyr UAA Stop UAG Stop	UGU UGC } Cys UGA Stop UGG Trp	U C A G
C	CUU CUC CUA CUG } Leu	CCU CCC CCA CCG } Pro	CAU CAC } His CAA CAG } Gln	CGU CGC CGA CGG } Arg	U C A G
A	AUU AUC AUA } Ile AUG Met/Start	ACU ACC ACA ACG } Thr	AAU AAC } Asn AAA AAG } Lys	AGU AGC } Ser AGA AGG } Arg	U C A G
G	GUU GUC GUA GUG } Val	GCU GCC GCA GCG } Ala	GAU GAC } Asp GAA GAG } Glu	GGU GGC GGA GGG } Gly	U C A G

Question 16 (7 marks) ©NESA 2016 SII Q34e ●●●

'Genes influence proteins and proteins influence genes.'

Evaluate this statement with reference to the structure and function of genes and proteins.

Test 4: Genetic variation

Section I: 10 marks. Section II: 28 marks. Total marks: 38.
Suggested time: 70 minutes

Section I: Multiple-choice questions

Instructions to students
- For each question, circle the multiple-choice letter to indicate your answer.

Question 1

In pea plants, there are two alleles for flower colour: the purple-flower allele (P), and the white-flower allele (p).

A pea plant has the genotype pp. Which of the following terms could be used to describe this pea plant?

A Hybrid

B Homozygous

C Heterozygous

D Purple-flowered

Question 2

Consider an organism with the genes A and B.

How many unique gametes could be produced by an individual with the genotype AaBB?

A 1

B 2

C 3

D 4

Question 3

Two plants with the same phenotype are crossed, resulting in a phenotypic ratio of 3 : 1 in the offspring.

It could be deduced that

A both parents are heterozygous.

B each of the offspring has the same genotype.

C one in every four offspring has the dominant phenotype.

D one of the alleles involved is incompletely dominant over the other.

Question 4

Marfan syndrome is a genetic disorder that affects the heart, blood vessels, bones and eyes. It is caused by a mutation in the *FBN1* gene. It is inherited as an autosomal dominant trait.

Based on the information above, which of the following is true regarding Marfan syndrome?

A It affects more males than females.

B It is expressed in heterozygous individuals.

C It is a common disorder, because of its dominant nature.

D A person with Marfan syndrome must have two mutated *FBN1* alleles.

Question 5 ©NESA 2020 SI Q14

A normal allele results in liver cells with sufficient cholesterol receptors. A different allele results in liver cells without cholesterol receptors. Individuals who are heterozygous have liver cells with insufficient cholesterol receptors.

What type of inheritance is the most likely explanation for this?

A Sex-linked

B Autosomal dominant

C Autosomal recessive

D Incomplete dominance

Use the following information to answer Questions 6 and 7.

Mallard ducks have a unique colour pattern that is controlled by multiple alleles: the mallard allele (M), the restricted allele (M^R) and the dusky allele (m^d).

The alleles display the following hierarchy of dominance:

$$M^R > M > m^d$$

Question 6

A male duck with the genotype M^Rm^d is mated with a female who is heterozygous for the mallard pattern.

The resulting offspring could have

A 3 genotypes and 2 phenotypes.

B 3 genotypes and 3 phenotypes.

C 4 genotypes and 3 phenotypes.

D 4 genotypes and 4 phenotypes.

Question 7

What percentage of the offspring would you expect to have the same phenotype as the male parent?

A 0%

B 25%

C 50%

D 100%

Question 8

Mitochondria contain their own genetic information, called mitochondrial DNA or mtDNA. Each mitochondrion contains a single, circular piece of mtDNA. As with nuclear DNA, changes in the mtDNA sequence may result in disease. Such conditions are referred to as mitochondrial diseases.

A human ovum contains between 100 000 and 600 000 mitochondria, while a human sperm contains between 50 and 75 mitochondria. Upon fertilisation, the paternal mitochondria enter the egg and usually are rapidly degraded by enzymes in the egg cytoplasm.

Which of the following shows the typical presence of a mitochondrial disease in a family?

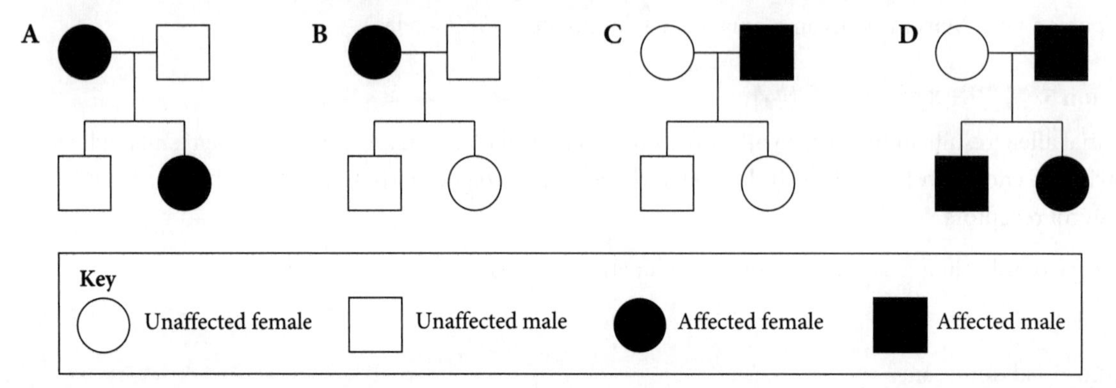

Question 9

Haplodiploidy is a sex-determination system in bees, ants and wasps. In these organisms, there are no distinct sex chromosomes. Instead, sex is determined by how many sets of chromosomes the individual has.

In haplodiploid organisms, gametes (eggs and sperm) are haploid. A female offspring is produced when fertilisation takes place, while a male offspring develops from an unfertilised egg.

Which of the following statements about haplodiploid organisms is correct?

A Males can produce both sons and daughters.

B Males have a grandfather and can have grandsons.

C Females have half the number of chromosomes that males have.

D Both males and females undergo meiosis to produce gametes.

Question 10 ©NESA 2021 SI Q18

Which human pedigree shows the inheritance of a recessive, sex-linked characteristic?

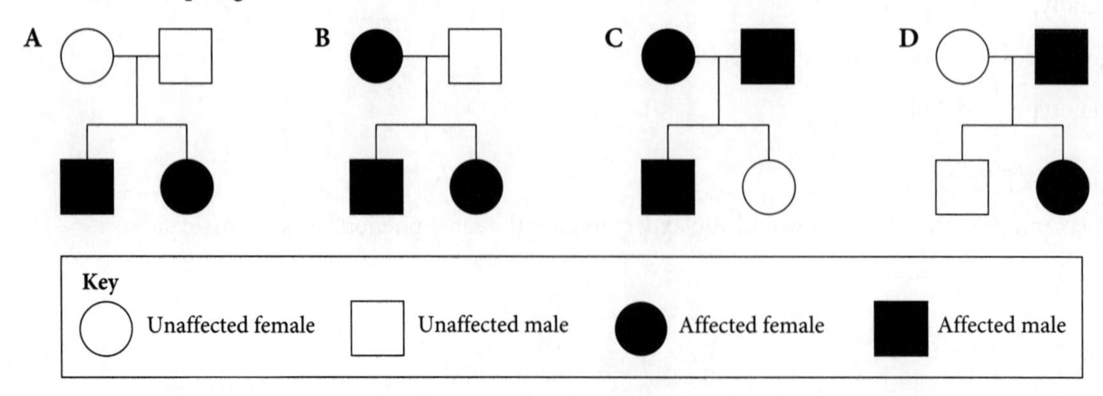

Section II: Short-answer questions

Instructions to students

- Answer all questions in the spaces provided.

Question 11 (4 marks)

Coat colour in cats exhibits codominance. When a tan cat and a black cat are mated, they produce offspring with 'tabby' coats, as shown in the Punnett square below.

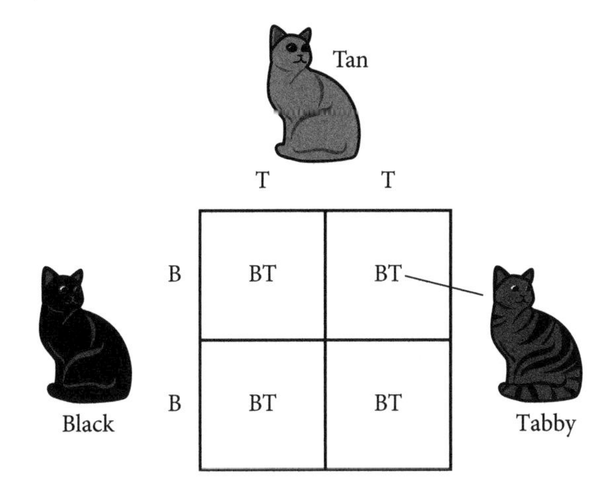

a If a tabby cat offspring from the Punnett square above was mated with a tan cat, what phenotypic ratio would be present in their offspring? 1 mark

b Using an example, explain how incomplete dominance differs from codominance. 3 marks

Question 12 (7 marks)

Tail docking is the process of surgically removing a portion of an animal's tail. The practice is now banned in Australia; however, many dog breeds have naturally bobbed (shortened) tails. These appear similar to the tails of docked dogs.

The autosomal condition, known as 'bob-tail' is often caused by a single nucleotide polymorphism in the *TBXT* gene.

a Define 'single nucleotide polymorphism'. 1 mark

b What is an autosomal condition? 1 mark

c The bob-tail (T) allele is dominant over the normal (t) allele. When two dominant alleles are present in an embryo, the pup will die in utero or shortly after birth. This phenotype is referred to as a 'lethal' phenotype.

The table below shows the phenotypes of a litter of puppies.

Phenotype	Number of pups
Lethal	1
Bob-tail	5
Normal tail	2

i Calculate the allele frequencies for this litter of puppies. 2 marks

Allele	Frequency
T	
t	

ii State the genotype of the parents involved in this cross. Justify your answer. 3 marks

Question 13 (5 marks) 

Menkes disease is a recessive sex-linked disorder caused by a mutation in the *ATP7A* gene. The diagram below shows inheritance of the disease in a family.

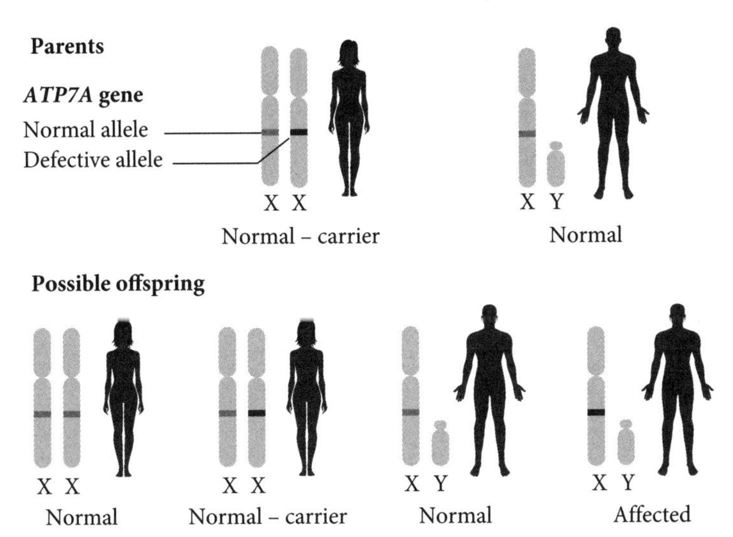

Explain why male offspring have a 50% chance of inheriting Menkes disease, while females have a 0% chance. Include a Punnett square in your answer.

Question 14 (6 marks)

Spinal muscular atrophy (SMA) is a genetic condition that causes muscle weakness and atrophy. There are four types of SMA, with each type exhibiting a different severity of symptoms and time of onset. In SMA type 4 (SMA 4), symptoms do not usually appear until the person is in their mid-30s.

Consider the following information about a family with a history of SMA 4.

> 'I am a female with SMA 4. I have two brothers: one has SMA 4, and the other does not. My mother and my father do not have the disease. My father's sister has SMA 4; however, his brother does not. My father's father does not have the disease, while his mother does. Neither of my mother's parents has the disease.'

a Construct a pedigree for this family in the space below. 3 marks

b Identify the mode of inheritance of this disorder. Justify your answer with reference to specific individuals. 3 marks

Question 15 (6 marks) ©NESA 2019 SII Q28

Huntington's disease is an autosomal dominant condition caused by a mutation of a gene on chromosome 4. It causes nerve cells to break down.

Stargardt disease is an autosomal recessive condition caused by a mutation of a different gene on chromosome 4. It causes damage to the retina.

A patient is heterozygous for both Huntington's (Hh) and Stargardt disease (Rr). His father's extended family has numerous cases of both of these diseases. His mother does not have either disease and is homozygous for both genes.

a Complete the tables, showing the **two** alleles the patient inherited from each parent. 2 marks

Alleles from father

Alleles from mother

b The diagram shows the patient's homologous pair of chromosome 4 at various stages of meiosis.

Add the relevant alleles to the diagram to model the production of possible gamete combinations. Include a key and an example of crossing over. 4 marks

Homologous pair of chromosome 4 before crossing over	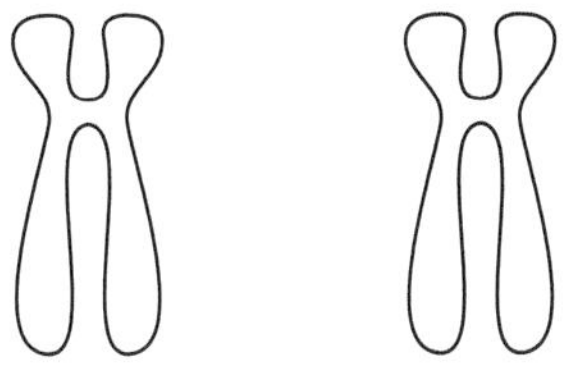	KEY
Homologous chromosomes after crossing over and separation		
Gametes	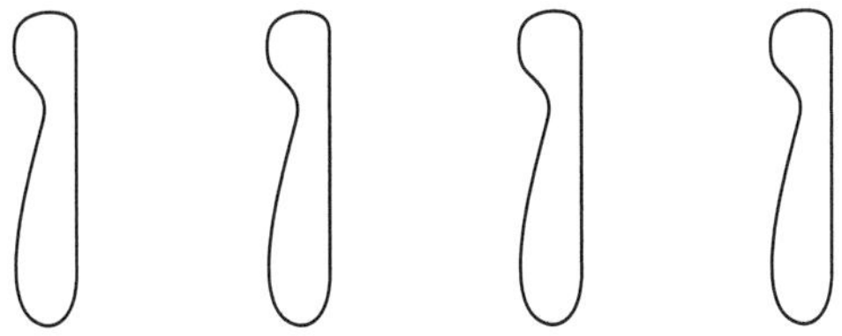	

Test 5: Inheritance patterns in a population

Section I: 10 marks. Section II: 30 marks. Total marks: 40.
Suggested time: 70 minutes

Section I: Multiple-choice questions

Instructions to students
- For each question, circle the multiple-choice letter to indicate your answer.

Question 1

What is a genome?

A The complete set of proteins within each cell

B The sequence of bases within the genes of each cell

C The complete set of genetic information within an organism

D The groups of genes expressed within each type of cell in an organism

Question 2

Which of the following is true regarding DNA sequencing?

A It can only be carried out on eukaryotic organisms.

B It can only be carried out on coding DNA segments.

C It can be used to determine the risk of inheriting a genetic disease.

D It can be used to amplify small amounts of DNA found at crime scenes.

Question 3

Rosellas are a genus of colourful parrots native to Australia. The genus consists of six species and many more sub-species, each of which differ in size, wing and tail shape, and plumage colour and pattern. Mating between closely related rosella species is common.

A scientist discovers a small rosella population on a remote island off the east coast of Australia.

Which of the following techniques would prove most useful in determining evolutionary relationships between this population and the species on the mainland?

A DNA sequencing

B Study of mating behaviours

C Analysis of amino acid sequences

D Study of morphological features such as size and colouring

Question 4

Fragments of DNA from extinct *Homo* species have been retrieved from museum specimens, archaeological finds and fossil remains. This DNA has been used to

A clone the ancient relatives of humans.

B study the evolutionary history of humans.

C reintroduce ancient human traits into living humans.

D prove that modern-day humans evolved from chimpanzees.

Use the following information to answer Questions 5 and 6.

Paternity testing involves a number of procedures used to determine the identity of a child's biological father. One such procedure is blood typing, which analyses the blood groups of the parties involved to determine the paternity.

There are four main blood groups in humans. The possible genotypes for each of these blood groups are shown in the table below.

Phenotype	Possible genotypes
Type A	I^AI^A or I^Ai^o
Type B	I^BI^B or I^Bi^o
Type AB	I^AI^B
Type O	i^oi^o

Another procedure used in paternity testing is DNA fingerprinting, in which gel electrophoresis is used to show the genetic patterns in an individual's DNA. These patterns are compared between the individuals concerned.

Question 5 ©NESA 2003 SII Q32c (ADAPTED)

The following data shows the results of a blood typing procedure attempting to identify the biological father of a child. Maternity of the child has been verified.

ABO blood groups

Mother	Child	Male 1	Male 2	Male 3
A	O	O	A	B

Based on the ABO blood group data, which male(s), if any, can be excluded as the father of this child?

A Male 1

B Male 2

C Male 3

D None of the males can be excluded.

Question 6 ©NESA 2003 SII Q32c (ADAPTED)

The following data shows the results of a DNA fingerprinting procedure attempting to identify the biological father of a child. Maternity of the child has been verified.

DNA fingerprint data

Molecular weight	Weight marker	Mother	Child	Male 1	Male 2	Male 3
6000	■				■	
5500	■	■	■			
5000	■					
3000	■	■			■	■
2000	■			■		
1500	■		■			■
1000	■			■		

Based on the DNA fingerprint data, which male is the father of this child?

A Male 1

B Male 2

C Male 3

D It is impossible to tell.

Question 7

Down syndrome is a genetic disorder caused by the presence of all or part of a third copy of chromosome 21. It is the most common chromosomal disorder in many countries, including Australia and the United States.

The graph shows the relationship between mother's age and the risk of Down syndrome in her child.

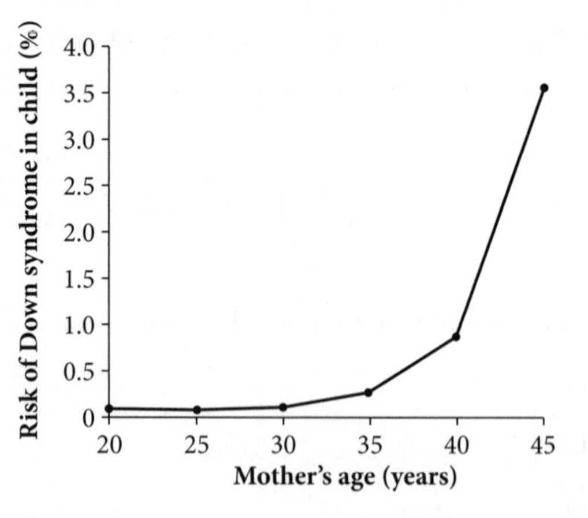

Source: D. Newberger, 'Down syndrome: Prenatal risk assessment and diagnosis', Figure 1, *American Family Physician*, 2000, 15 Aug; 62(4): 825–832)

In the United States between 1979 and 2003, the number of babies born with Down syndrome increased by about 30%.

Which of the following would account for this increased incidence?

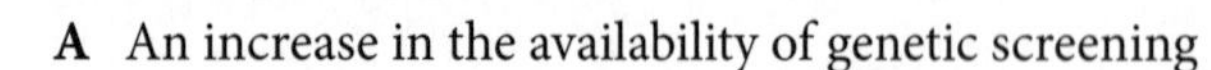

A An increase in the availability of genetic screening

B An increase in the age at which women conceived

C An increase in the life expectancy of people with Down syndrome

D An increase in the termination rate in instances where Down syndrome was diagnosed during pregnancy

Question 8

Single nucleotide polymorphisms (SNPs) occur when a single nucleotide in the DNA sequence is altered from one individual to the next. Genome-wide association studies (GWAS) compare the frequency of certain SNPs in healthy and diseased populations to identify SNPs that are associated with increased risk of disease. Over 90% of the SNPs identified in GWAS are found in non-coding regions of the DNA.

How might a SNP in a non-coding region increase the risk of a disease?

A It might cause changes in gene expression.

B It might change the instructions for making a protein.

C It might lead to a single nucleotide change in mature mRNA.

D It might lead to a change in the amino acid sequence of an important polypeptide.

Question 9

The diagram below shows a section of DNA. A particular restriction enzyme cuts the DNA at two points, producing fragments X, Y and Z.

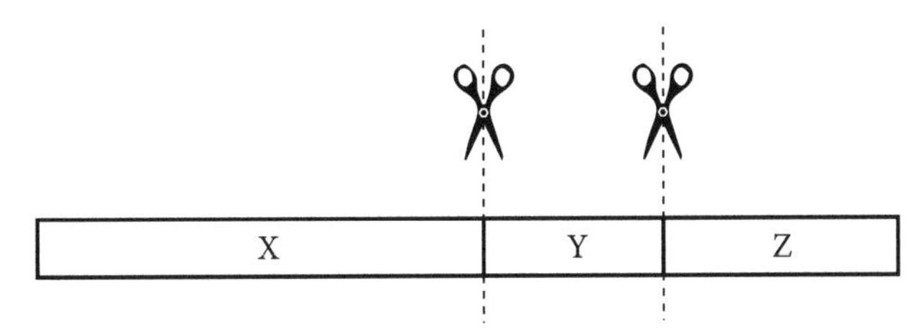

Which of the following most closely represents the banding pattern that would appear if the fragments were separated by gel electrophoresis?

A

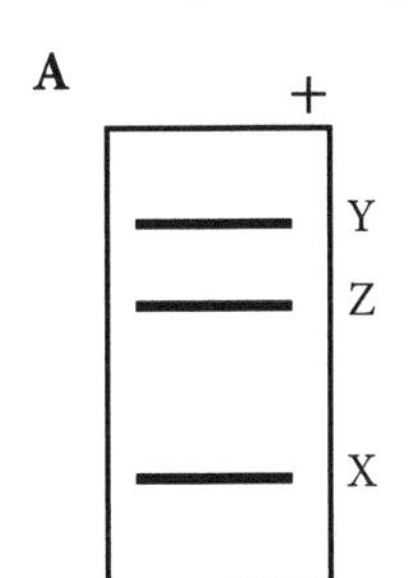

B

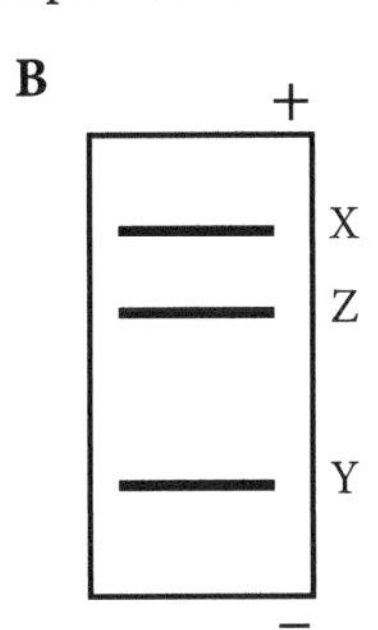

C

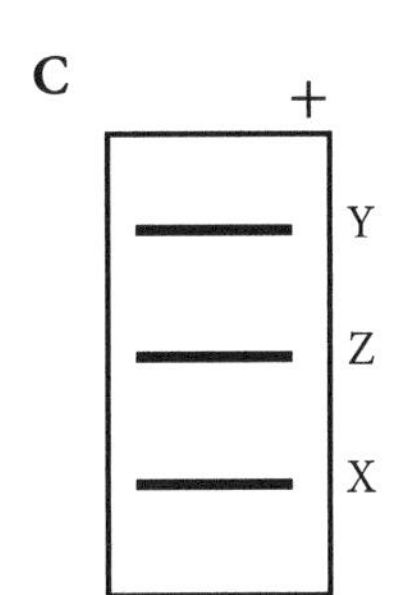

D

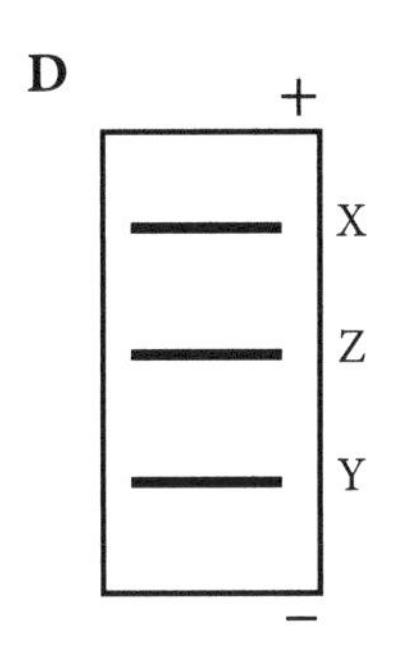

Question 10

Humans have 23 pairs of chromosomes, whereas non-human primates have 24 pairs. There is great similarity between human chromosomes and those of non-human primates, except for human chromosome pair number 2.

One of the chromosomes from this pair in humans is shown on the right, along with the equivalent chromosomes in chimpanzees and gorillas. The light and dark bands on the chromosomes represent the location of genes.

Human

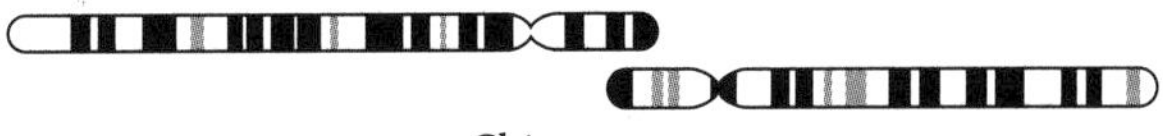

Chimpanzee

Gorilla

What conclusion about human evolution can be drawn from this data?

A Humans evolved from chimpanzees.

B Humans are more closely related to gorillas than chimpanzees.

C Non-human primates evolved from humans when one pair of chromosomes broke in half.

D Human chromosome 2 is the result of the fusion of two chromosomes in the common ancestor of all primates.

Section II: Short-answer questions

Instructions to students
- Answer all questions in the spaces provided.

Question 11 (4 marks)

DNA profiling is a technology by which individuals can be identified and compared according to their DNA. It uses a number of small sections of a person's DNA, which are sorted according to size, to build a DNA 'fingerprint'. DNA fingerprints can be used to determine patterns of disease inheritance, paternity and immigration eligibility, and as evidence in forensic investigations.

The diagram below shows the DNA fingerprints from a forensic investigation.

Evidence Victim Suspect 1 Suspect 2 Suspect 3

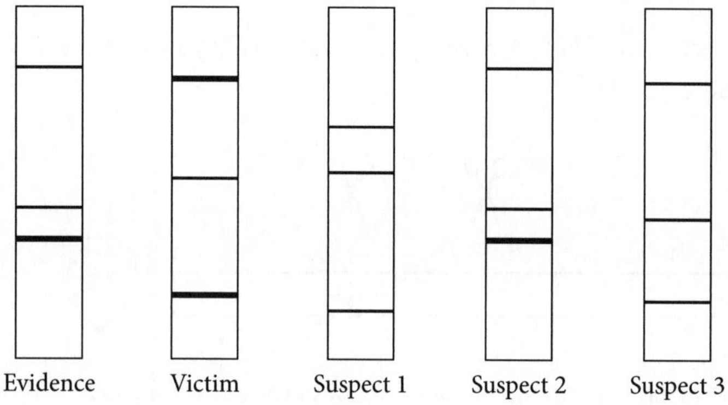

a Identify the suspect who is likely to be the perpetrator. Justify your choice. 2 marks

b Explain why a sample was taken from the victim as well as the suspects. 2 marks

Question 12 (5 marks)

Mitochondria contain a single, small circular piece of DNA, called mtDNA. Mitochondria are present in both egg and sperm; however, when the paternal mitochondria enter the egg during fertilisation, they are usually rapidly degraded by enzymes in the egg cytoplasm, so that only maternal mitochondria remain.

The diagram shows how nuclear DNA and mtDNA are passed down from generation to generation.

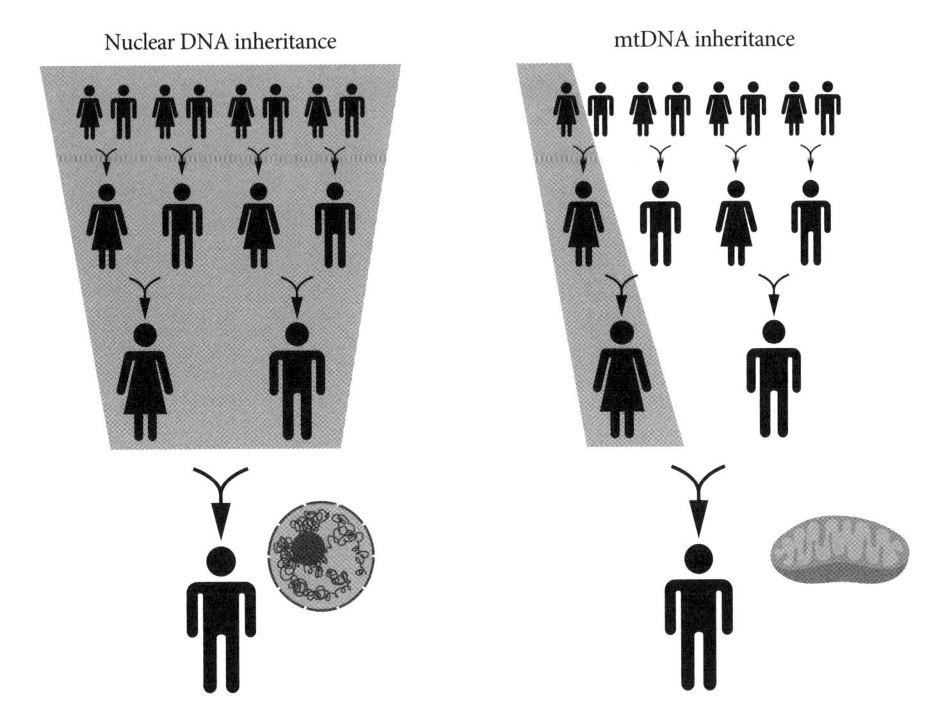

a Compare the pattern of inheritance of nuclear DNA with that of mtDNA. 1 mark

b mtDNA is often more useful than nuclear DNA when studying population genetics relating to human evolution. Explain **two** possible reasons for this. 4 marks

Question 13 (3 marks) ©NESA 2021 SII Q27

Sickle cell anaemia is a genetic disorder. In one family, the parents are both known to be heterozygous for the mutation that causes sickle cell anaemia. The couple have two unaffected children and are expecting a third child. They have had an allele screening test to determine whether the child will have sickle cell anaemia.

A part of the DNA profile is shown. It shows the alleles present.

Mother	Father	Child 1	Child 2	Child 3
▬	▬		▬	▬
▬	▬	▬	▬	

Use the DNA profile provided to justify whether Child 3 will have sickle cell anaemia.

Question 14 (8 marks)

Huntington's disease (HD) is a rare neurodegenerative disorder that is inherited in an autosomal dominant fashion.

The HD gene contains a series of repeating CAG nucleotides. People who have fewer than 27 CAG repeats are normal, while those with 40 or more will have Huntington's disease. Having between 27 and 39 repeats may not result in disease expression, but may result in expression in offspring, if they inherit the affected allele.

The number of CAG repeats and the median age of onset are shown in the table.

CAG repeat size	Median age of onset (years)
40	59
45	37
50	27
55	23
60	20

a Plot the data on the grid provided. 4 marks

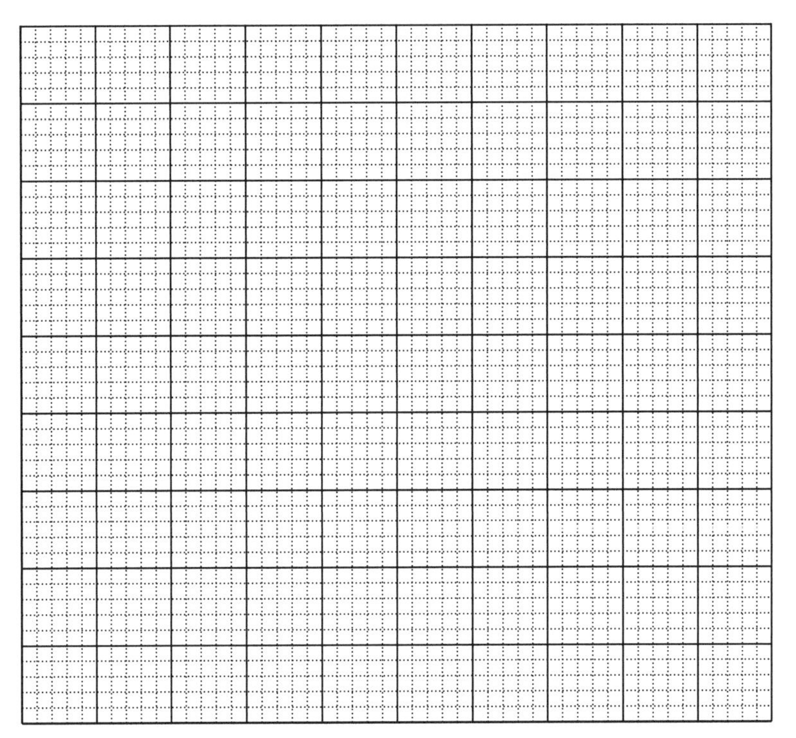

b Describe the trend shown in the data. 2 marks

c Explain why collaboration on a large scale is needed to gather reliable data on the inheritance of rare diseases such as Huntington's disease. 2 marks

Question 15 (10 marks)

Cystic fibrosis (CF) is caused by a range of mutations in the *CFTR* gene. Homozygous recessive individuals have defective chloride channels in their epithelial cells, resulting in thick, sticky mucus building up in the lungs and gut. This leads to progressive and often fatal lung damage. Heterozygous individuals have some defective chloride channels, but not enough to cause symptoms of CF.

The incidence of CF in several ethnic groups, along with the frequency of the heterozygous genotype, is shown in the table.

Ethnic group	Incidence of CF (per 100 000 population)	Frequency of heterozygous genotype
Caucasian	31.25	1 in 29
Hispanic	8.33	1 in 46
African	6.67	1 in 65
Asian	2.85	1 in 90

a What is the relationship between the frequency of the heterozygous CF genotype and the incidence of CF? 1 mark

b The graph below shows data relating to the effect of CF alleles on the ability of pathogenic bacterial strains to enter the cells of the gut epithelium.

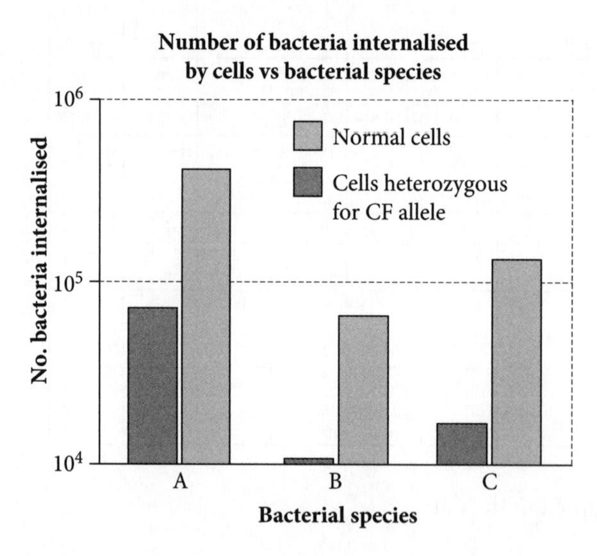

i What conclusion can be drawn from this data? 1 mark

ii ●●● Based on the information provided throughout this question, propose an explanation for the increased frequency of the heterozygous CF genotype in Caucasian populations. 4 marks

c ●●● Since the discovery of the CFTR gene in 1989, a group of 130 CF laboratories (from more than 30 countries) have been working together to identify and catalogue the growing number of CF-causing mutations.

Explain **two** benefits of large-scale collaboration when identifying trends, patterns and relationships in the inheritance of genetic diseases such as CF. 4 marks

CHAPTER 2
MODULE 6: GENETIC CHANGE

Test 6: Mutation

Section I: 10 marks. Section II: 30 marks. Total marks: 40.
Suggested time: 70 minutes

Section I: Multiple-choice questions

Instructions to students

- For each question, circle the multiple-choice letter to indicate your answer.

Question 1

A newborn baby was diagnosed with Edwards syndrome. Karyotyping revealed he had three copies of chromosome 18.

What type of mutation is the one that causes Edwards syndrome?

A Polyploidy

B Aneuploidy

C Point mutation

D Insertion mutation

Question 2

A mutation that does not cause a change in an amino acid is called a

A silent mutation.

B missense mutation.

C nonsense mutation.

D frameshift mutation.

Question 3 ©NESA 2021 SI Q6

A mutation involving a DNA deletion is illustrated.

Which statement about the mutation is correct?

A It will have an effect on many genes.

B It will have an effect on only one codon.

C It may be the result of an error during translation.

D It may be the result of an error during transcription.

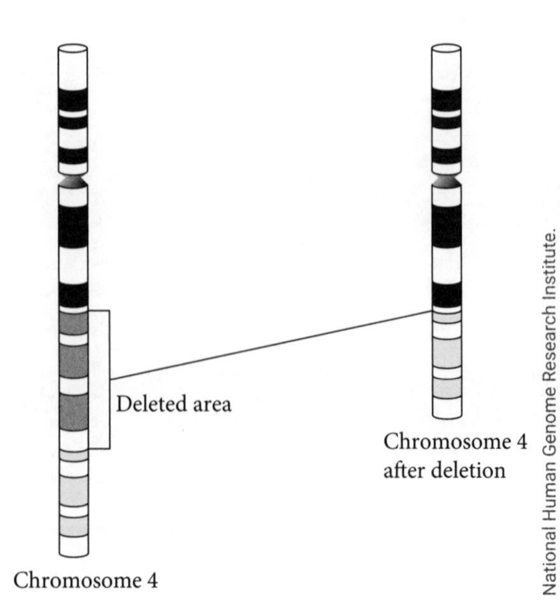

National Human Genome Research Institute.
https://www.genome.gov/genetics-glossary/deletion

Question 4

The diagram below shows a model of the bottleneck effect, which occurs when the size of a population is suddenly and dramatically reduced as a result of a catastrophic event.

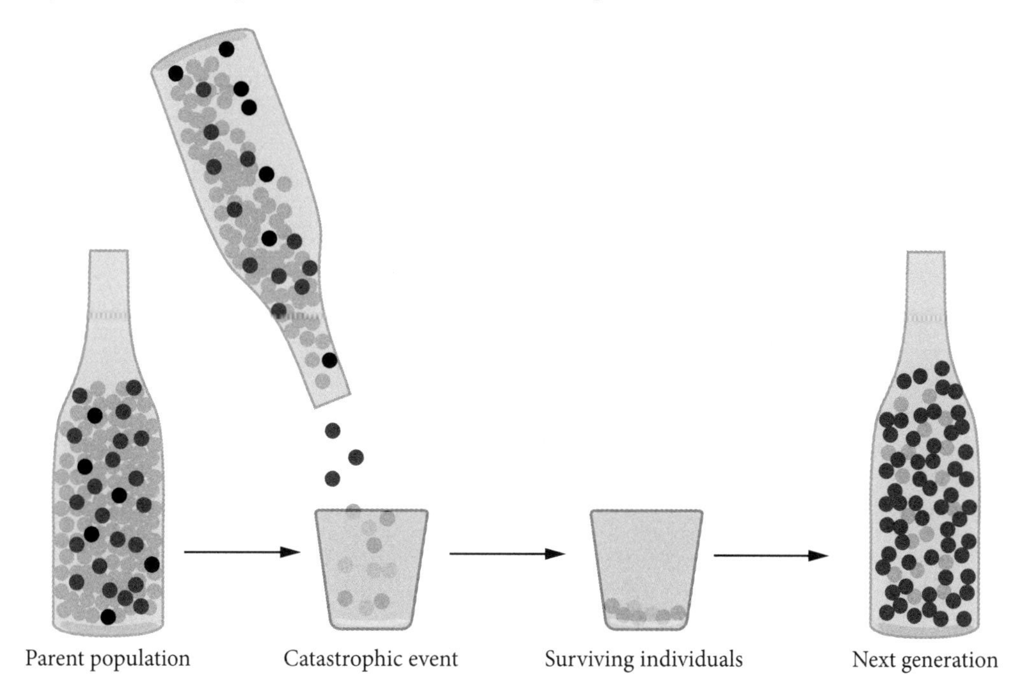

The bottleneck effect is an example of

A mutation.

B gene flow.

C genetic drift.

D natural selection.

Question 5

Which of the following statements about non-coding DNA segments is true?

A They are not affected by mutagens.

B They are expressed as polypeptides.

C They are found both between and within genes.

D They make up only a very small percentage of an individual's genome.

Question 6

No two people are genetically identical, except for identical twins. The main cause of genetic variation between two human individuals is

A environmental factors that influence phenotype.

B new mutations that occurred in the previous generation.

C genetic drift because of the small size of the population.

D the reshuffling of alleles as a result of sexual reproduction.

Use the information below to answer Questions 7 and 8.

In humans, the Y chromosome carries the sex-determining region Y (*SRY*) gene. The *SRY* gene provides instructions for making a protein that causes a foetus to develop testes. In the absence of an *SRY* gene, the foetus will develop female reproductive organs.

In rare cases, a mutation may result in the *SRY* gene being found on the X chromosome, as shown in the diagram below.

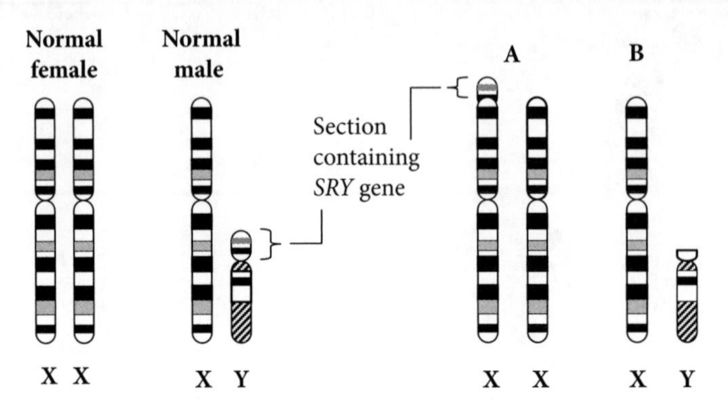

Question 7

What type of mutation is shown in the diagram?

A Aneuploidy

B Translocation

C Point mutation

D Frameshift insertion

Question 8

What effect would this mutation have on the sex of individuals A and B?

	Individual A	Individual B
A	Female	Female
B	Female	Male
C	Male	Male
D	Male	Female

Question 9

The molecule 5-bromouracil is similar in structure to thymine; DNA polymerase cannot distinguish between them. In some cases, 5-bromouracil pairs correctly with adenine, but it can also pair with guanine.

Why is 5-bromouracil considered to be a mutagen?

A It causes a kink in the DNA backbone.

B It causes chromosomal rearrangements.

C It causes base substitutions during DNA replication.

D It causes a change in the reading frame of the ribosome.

Question 10 ©NESA 2020 SI Q17

There are about 10 million single nucleotide polymorphisms (SNPs) in the human genome. Four SNPs are modelled in the diagram.

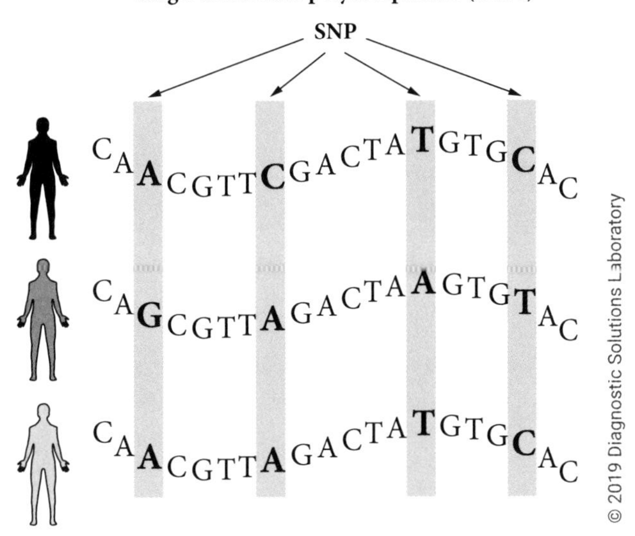

The SNPs modelled do not affect the phenotype of the individuals shown.

Which is the best explanation for this?

A Only one nucleotide is different at each SNP.

B The SNPs are part of DNA that is not expressed.

C AGA, CAA, TAT and CTC all code for the same amino acid.

D The SNPs are present on one strand of the DNA molecule only.

Section II: Short-answer questions

Instructions to students
- Answer all questions in the spaces provided.

Question 11 (3 marks)

a Define 'mutagen'. 1 mark

b Identify a mutagenic electromagnetic radiation source and outline how it operates as a mutagen. 2 marks

Question 12 (4 marks)

Turner syndrome is a chromosomal disorder that affects approximately one in 2000 females. Women with Turner syndrome typically have a short stature, absence of menstruation and infertility.

The karyotype of a woman with Turner syndrome is shown below.

Turner syndrome

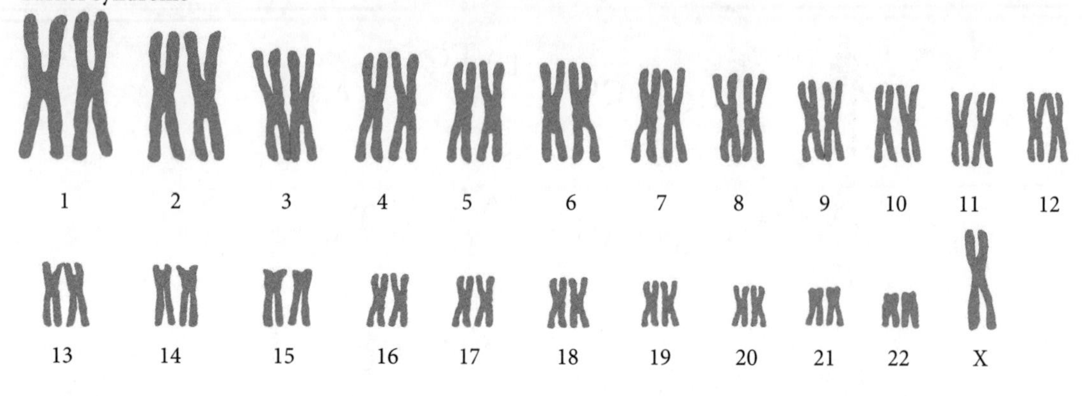

a Identify the cause of Turner syndrome. 1 mark

b Explain how mutation during meiosis could lead to an offspring with Turner syndrome. 3 marks

Question 13 (7 marks) ©NESA 2021 SII Q24

An incidence of an autosomal dominant trait is shown in the pedigree.

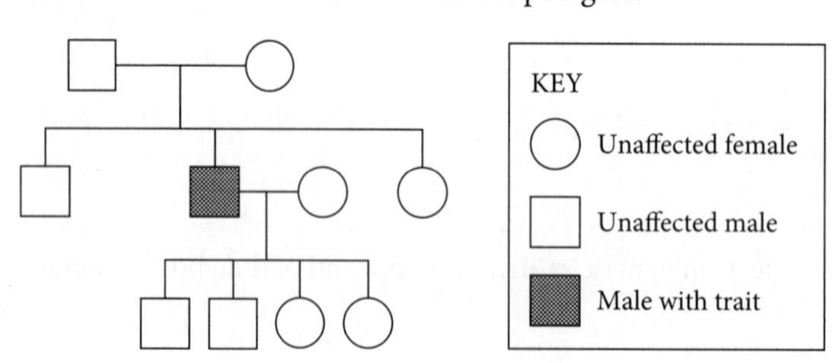

a Is this trait likely to be the result of a somatic or a germ-line mutation? Justify your answer. 3 marks

b The diagram shows the early stages of embryonic development from a fertilised egg. The developing ball of cells has split and monozygotic (identical) twins have formed. Mutations can occur at different times during embryonic development, for example Mutation *A* would result in both twins having the mutation in all their cells.

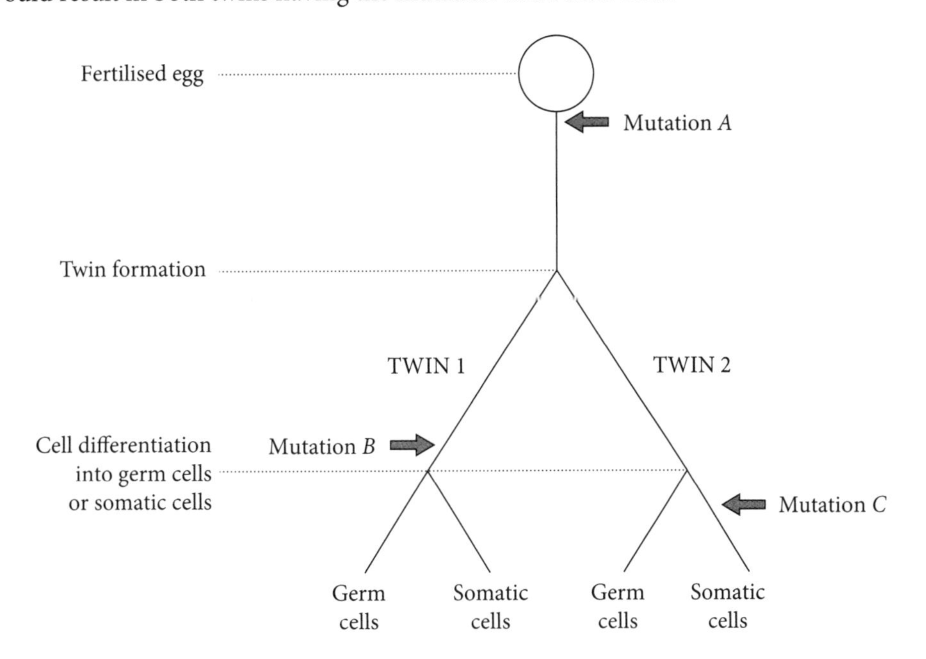

Explain the effects of Mutation *B* and Mutation *C* on each twin and on any offspring that they may have. 4 marks

Question 14 (8 marks)

Tay–Sachs disease is a rare genetic condition caused by mutations in the *HEXA* gene. This gene, which is expressed in cells of the brain and spinal cord, provides instructions for making an enzyme called beta-hexosaminidase A.

This enzyme helps to break down a fatty substance called GM2 ganglioside. Build-up of GM2 ganglioside leads to the destruction of neurons in the brain and spinal cord. This leads to the signs and symptoms of Tay–Sachs disease, which include muscle weakness, seizures, vision loss and, eventually, paralysis.

The table below shows a section of template DNA from a normal HEXA allele. The most common Tay–Sachs HEXA allele is also shown. The mutation is shown in bold.

Normal HEXA allele	GCA	TAT	AGG	ATA	CGG	GGA	CTG	…	
Tay–Sachs HEXA allele	GCA	TAT	AG**A**	**TAG**	GAT	ACG	GGG	ACT	G…

a Identify the type of mutation that causes Tay–Sachs disease. 1 mark

b The table below shows the codons in mRNA and their corresponding amino acid.

Second base (columns) / **First base** (rows) / **Third base** (right)

First base	U	C	A	G	Third base
U	UUU, UUC: Phe UUA, UUG: Leu	UCU, UCC, UCA, UCG: Ser	UAU, UAC: Tyr UAA Stop UAG Stop	UGU, UGC: Cys UGA Stop UGG Trp	U C A G
C	CUU, CUC, CUA, CUG: Leu	CCU, CCC, CCA, CCG: Pro	CAU, CAC: His CAA, CAG: Gln	CGU, CGC, CGA, CGG: Arg	U C A G
A	AUU, AUC, AUA: Ile AUG Met/Start	ACU, ACC, ACA, ACG: Thr	AAU, AAC: Asn AAA, AAG: Lys	AGU, AGC: Ser AGA, AGG: Arg	U C A G
G	GUU, GUC, GUA, GUG: Val	GCU, GCC, GCA, GCG: Ala	GAU, GAC: Asp GAA, GAG: Glu	GGU, GGC, GGA, GGG: Gly	U C A G

Complete the tables below to show how transcription and translation are affected by the mutation. 4 marks

Normal HEXA DNA	GCA	TAT	AGG	ATA	CGG	GGA	CTG	…
Normal HEXA mRNA								
Normal HEXA polypeptide								

Tay–Sachs HEXA DNA	GCA	TAT	AG**A**	**TAG**	GAT	ACG	GGG	ACT	G…
Tay–Sachs HEXA mRNA									
Tay–Sachs HEXA polypeptide									

c Explain how this mutation would lead to the symptoms of Tay–Sachs disease. 3 marks

Question 15 (8 marks)

Numbats (*Myrmecobius fasciatus*) are small marsupials found in southern Western Australia. They were once widespread in several states but are now considered endangered, and it is estimated that there are currently fewer than 1000 mature numbats in the wild. The biggest threats to numbats are predation by feral cats and foxes, inappropriate hazard reduction burning, and habitat disturbance from farming and timber harvesting.

Only two natural populations of numbats remain, one in the Dryandra Woodland and another in the Upper Warren area. Numbats have been taken from these populations and translocated to a number of sites within the former range of the species, in an effort to increase wild population numbers.

In 1985, thirty-five founder animals from the Dryandra population were released at Boyagin Nature Reserve to form a subpopulation. Four more Dryandra numbats have been added since the original translocation (a male and a female in 2005, and a male and a female in 2010). Breeding has been recorded and the subpopulation is considered to be self-sustaining, with an estimated 50–100 individuals.

The map below shows the location of the two remaining natural populations, as well as a number of subpopulations that have been established in southern Western Australia as a result of translocations. Each population and subpopulation are genetically isolated.

Map of southern Western Australia

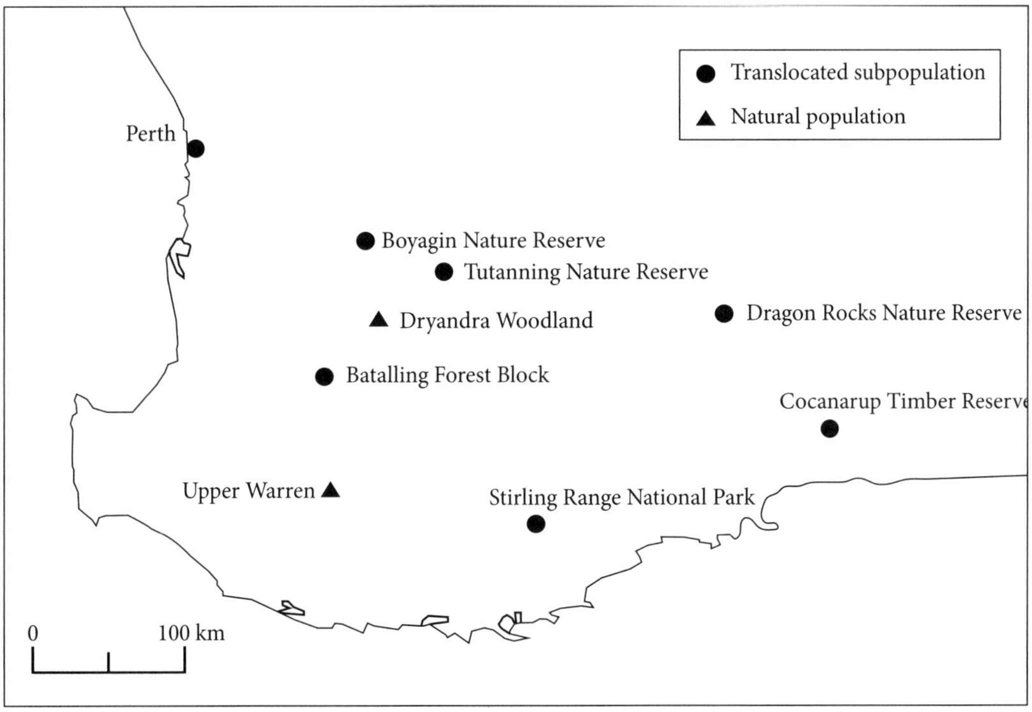

a Define 'gene pool'. 1 mark

b ●●● Evaluate the effects of mutation, gene flow and genetic drift on the gene pool of the numbat populations. 7 marks

Test 7: Biotechnology

Section I: 10 marks. Section II: 26 marks. Total marks: 36.
Suggested time: 65 minutes

Section I: Multiple-choice questions

Instructions to students
- For each question, circle the multiple-choice letter to indicate your answer.

Question 1

Which of the following is **not** an example of a biotechnology?

A The development of vaccines using weakened bacteria

B The fermentation of foods such as yoghurt using *Lactobacillus*

C The extraction of antibiotics such as penicillin from *Penicillium* mould

D The removal of weeds by humans to reduce their impact on native species

Question 2

Which of the following describes biodiversity at the species level?

A The variety of genes with a population

B The variety of ecosystems in a given place

C The variety of bases within an individual's DNA

D The variety of organisms within a habitat or region

Use the following information to answer Questions 3 and 4.

Potato plants are typically cultivated by allowing a potato to sprout, then cutting the potato into pieces, which are dried briefly, then planted. The shoots give rise to new plants.

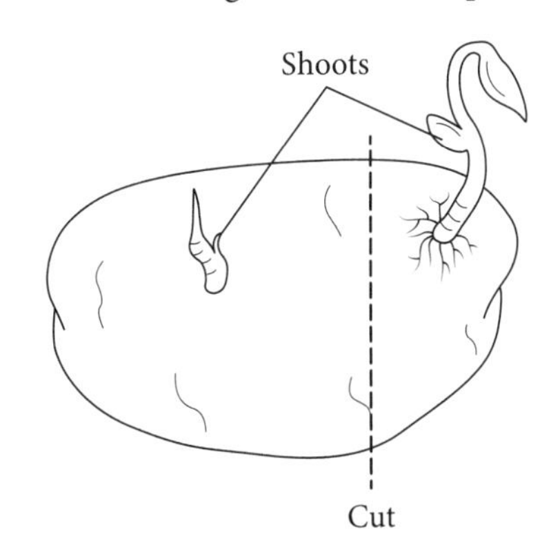

Question 3

The cultivation of potatoes in this way is an example of which biotechnology?

A Cloning

B Hybridisation

C Selective breeding

D Artificial pollination

Question 4

The potato was initially introduced to Ireland in the early 1800s to meet the needs of a growing population. It was cheap to grow and a good source of nutrition. Almost all farmers exclusively grew the Irish Lumper variety, because it had a high yield and tolerated moist soils. In 1845, a fungus called *Phytophthora infestans* decimated Ireland's potato crops, causing a famine that killed 1 million people.

Which of the following was the main cause of the Irish Potato Famine?

A Mutation

B Genetic drift

C Lack of biodiversity

D Overuse of insecticides

Question 5 ©NESA 2021 SI Q11

Many transgenic crops have been genetically engineered to have traits such as herbicide resistance. In at least four different crops the transgene has been found in nearby wild plant relatives of the cultivated crops.

What is the most likely reason for this observation?

A Crossing over in the wild plants

B Gene flow from the crops to the wild plants

C Genetic drift from the crops to the wild plants

D Mutations in the wild plants that match the transgenes

Question 6

Omega-3 oils are polyunsaturated fatty acids that are essential for brain and eye development. They can also reduce the risk of coronary heart disease, stroke, type 2 diabetes and Alzheimer's disease. As awareness of these health benefits has grown, so has our demand for omega-3 oils, with fish being the world's primary source.

In 2018, a new genetically modified (GM) strain of canola was approved for use in Australia. The GM canola contains high levels of omega-3 oils, with one hectare of GM canola producing the same amount of omega-3 as 10 tonnes of fish. Research studies have shown that the GM canola maintains the nutritional profile of traditional canola, and is safe for human and animal consumption.

What is a potential benefit for society of the commercial use of this GM canola?

A It could be used to cure diseases such as type 2 diabetes.

B It could reduce the amount of land required for canola farming.

C It could relieve pressure on wild fish stocks as a source of omega-3 oils.

D It would have no adverse effect on biodiversity, as it does not involve selective breeding.

Question 7

A team of researchers are discussing the merits of a hormone treatment used to treat menopause. The hormone is isolated from the oestrogen-rich urine of female horses. The researchers are discussing whether the benefits for the women being treated outweigh the harmful consequences for the horses.

Which of the following does the researchers' discussion address?

A Economic impacts

B Social implications

C Cultural influences

D Ethical considerations

Question 8

Rice is a staple food for more than 3.5 billion people around the world. The International Rice Research Institute in the Philippines maintains the world's largest bank of frozen rice seeds. Although it is home to more than 130 000 rice varieties, only a small number these of strains are currently used in commercial food production.

The primary reason for maintaining a bank of rice varieties is to

A store all strains until conditions for growing rice improve.

B reduce the genetic diversity of rice growing in the wild.

C prevent farmers from developing hybrids of wild strains without approval.

D maintain strains with characteristics that may prove useful for future commercial varieties.

Question 9

In 2022, a man in the US became the first person in the world to receive a heart transplant from a genetically modified pig. The pig used in the transplant had been genetically modified to 'knock out' several genes that would have led to the organ being rejected by the patient's immune system. To reduce the risk of infection, the donor pig was raised in a disease-free, laboratory environment and screened for many known pig pathogens before being brought to the laboratory.

Which of the following is a potential benefit of this technology for society?

A It will dramatically reduce the cost of heart transplants.

B It will eliminate the risk of infection for heart transplant recipients.

C It will eliminate the ethical concerns associated with receiving donated organs.

D It will reduce the number of people who die while waiting for an organ transplant.

Question 10 ©NESA 2016 SI Q18

How does the production of a new transgenic species have the potential to alter the path of evolution?

A The creation of new genes increases biodiversity.

B The removal of genes from a species decreases biodiversity.

C The transfer of genes within a species increases biodiversity.

D The transfer of genes between two species increases biodiversity.

Section II: Short-answer questions

Instructions to students
- Answer all questions in the spaces provided.

Question 11 (4 marks)

a Define 'biotechnology'. 1 mark

b Outline one example of a past biotechnology and explain how it meets the definition you gave in part **a**. 3 marks

Question 12 (4 marks)

Papaya is Hawaii's second most important fruit crop. Historically, artificial pollination was used to develop papaya varieties with excellent flavour and appearance. In the 1990s, Hawaiian papaya plantations were decimated by the uncontrollable spread of the papaya ringspot virus. Genetically modified (GM) papayas were developed to be resistant to the virus by adding a gene from the virus itself. Today, GM papayas account for approximately 90% of Hawaii's papaya production.

Explain how the **two** genetic techniques identified above affected the genetic diversity within Hawaii's papaya populations.

Question 13 (5 marks) ©NESA 2010 SII Q32c

A farmer raised some animals. Offspring from each generation were chosen for crosses for the next generation. The farmer collected data which are shown in the table.

Generation number	Breeding males	No. of offspring	No. of early deaths	Number of male animals surviving to breeding age
1	non-brown	53	27	2 non-brown 8 brown
2	brown	45	12	3 non-brown 15 brown
3	brown	62	10	1 non-brown 27 brown

a Outline the form of biotechnology used by the farmer. 1 mark

b Using the farmer's data, assess the effectiveness of this use of biotechnology. 4 marks

Question 14 (6 marks) 

GM crops to the rescue

It is estimated that global food production will need to double by 2050 to feed the world's growing population. Scientists believe that genetically modified (GM) crops, such as sorghum and eggplant, will play a major role in addressing global food shortages.

Sorghum is a drought- and heat-tolerant grain. It is a staple food for around 500 million people in low-income countries and is widely used as an animal feed for pigs and poultry. Researchers have produced transgenic sorghum, which has larger grains with increased protein content and digestibility. These transgenic lines have seen limited use because the ease of pollen dispersal means that they can easily hybridise with wild relatives to produce noxious weeds. Farmers looking to grow the transgenic lines are concerned that they will be required to enter into an agreement with the technology provider to purchase new seed every year.

Eggplant is a staple crop in India and an important source of income for many farmers. Every year, around 50% of crops are lost as a result of a pest called the eggplant fruit and shoot borer. Scientists have developed a genetically modified eggplant using the Bt gene from the bacterium *Bacillus thuringiensis.* The Bt eggplant produces a protein that kills borers. Since its introduction, farmers have reported a three-fold increase in yield, and have reduced pesticide use by as much as 92%. Bt eggplant crops must be carefully managed to avoid the evolution of insect resistance to the Bt toxin.

Use the information in the article above to complete the table below.

	GM sorghum	Bt eggplant
Potential benefit to society		
Social implications		
Future direction for research		

Question 15 (7 marks)

The theory of utilitarian ethics states that the most ethical choice is the one that will produce the greatest balance of good over harm for the greatest number. It judges an act based on the consequences of that act.

The extract below describes recent research involving genetic technology.

Transgenic mice model Alzheimer's disease

Alzheimer's disease (AD) is the leading cause of dementia. In 2021, an estimated 45 million people worldwide were living with AD. It is a progressive disease and there is currently no effective treatment. Declining memory, disorientation, and problems with speech and concentration mean that people with advanced AD need a high level of care. The estimated total healthcare cost of treating AD in 2020 was $305 billion.

Since the late 1980s, scientists have discovered several genes which, when mutated, appear to be linked to an increased risk of AD. More than 170 types of transgenic mice that express these mutant genes have been produced. Like a human with AD, a transgenic mouse develops plaques and 'tangles' in their brain, leading to significant, age-dependent cognitive impairment. Because these symptoms in the transgenic mice mimic the disease in humans, these mice can be used to study disease progression in humans, and to evaluate potential therapeutic strategies and drug candidates.

Adapted from: Elder G.A., Gama Sosa M.A. & De Gasperi R. (2010) Transgenic mouse models of Alzheimer's disease. *Mount Sinai Journal of Medicine*, 77(1): 69–81. https://doi.org/10.1002/msj.20159

Using the theory of utilitarian ethics, discuss how the development of new biotechnologies has led to ethical issues. Refer to the information in the passage in your answer.

Test 8: Genetic technologies

Section I: 10 marks. Section II: 28 marks. Total marks: 38.
Suggested time: 70 minutes

Section I: Multiple-choice questions

Instructions to students
- For each question, circle the multiple-choice letter to indicate your answer.

Question 1 ©NESA 2021 SI Q3

A scientist transferred male gametes from one plant to another to achieve a desired characteristic in the offspring. Which genetic technology was the scientist using?

A Gene cloning

B Artificial pollination

C Artificial insemination

D Whole-organism cloning

Question 2

Over the past 2000 years, humans have produced many modern vegetables from the wild mustard, *Brassica oleracea*. The wild plant and some of the most popular varieties are shown in the diagram below.

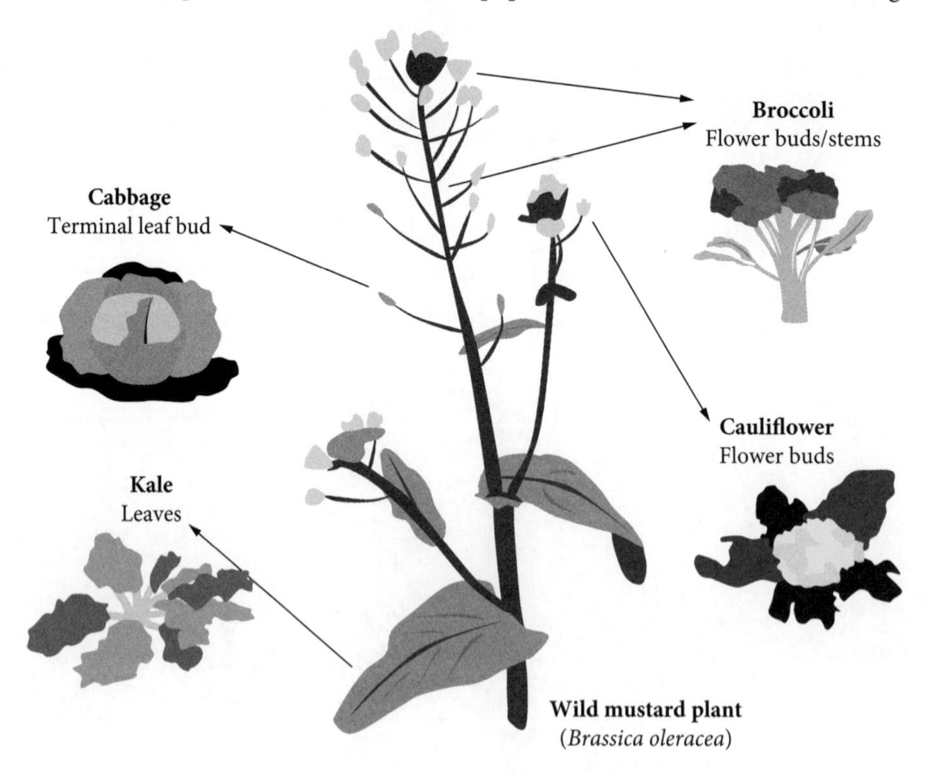

How have humans created such a variety of vegetables from one plant species?

A Natural selection was used to adapt the plant to the needs of humans.

B Different genes were added to the wild mustard plant to produce offspring with different traits.

C Parent plants with specific traits, such as larger buds, were bred together over many generations.

D The wild mustard plant was grown in different environments and naturally evolved different traits.

Question 3

Which of the following techniques is shown in the diagram below?

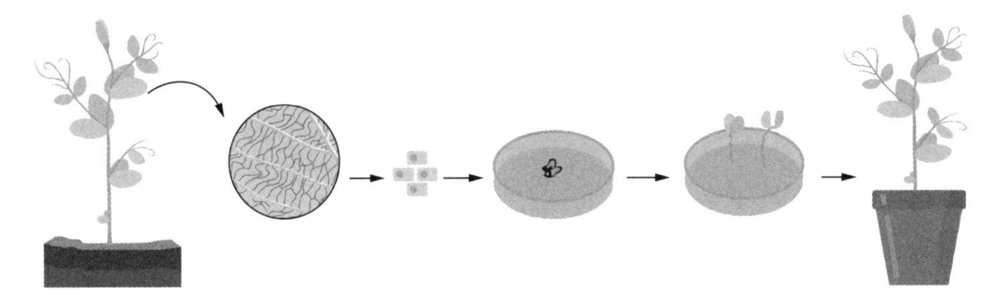

A Transgenics

B Gene cloning

C Tissue culture

D Artificial pollination

Question 4

Hybridisation within a species is the process of crossing individuals from two genetically distinct populations of the same species to produce a hybrid.

Which of the following is true regarding this type of hybridisation?

A It is a form of selective breeding.

B It produces offspring that are sterile.

C It decreases the genetic variation of the two populations.

D It results in offspring with a different chromosome number to the parents.

Question 5 ©NESA 2018 SI Q10

Both artificial insemination and cloning are reproductive techniques that can decrease the genetic diversity of a population.

Which row of the table provides a correct reason for each technique's contribution to this decrease?

	Artificial insemination	Cloning
A	Random fertilisation takes place	Large numbers of individuals are produced
B	One male has many offspring	All male gametes are identical
C	All male gametes are identical	All individuals have the same phenotype
D	Fewer males are used to reproduce	All individuals have the same genotype

Question 6

Artificial pollination may be carried out by hand or by machine. What is a benefit of hand pollination?

A It improves the chance of successful fertilisation.

B It is less time consuming than machine pollination.

C It produces many offspring, which are genetically identical.

D It introduces genes from other species, which may be beneficial.

Question 7

Which of the following is an example of transgenics?

A A labrador and a poodle are crossed to produce a labradoodle.

B Genes that cause cancer are silenced in ovarian cancer cells, leading to a reduction in tumour size.

C Instructions for the production of the bacterial enzyme phytase are incorporated into the genetic material of pigs.

D Wheat embryos are exposed to the chemical mutagen colchicine, to induce polyploidy and improve grain yield.

Question 8

Mutation induction involves intentionally exposing organisms to physical or chemical mutagens.

Why would mutation induction be useful in agriculture?

A It slows down the process of evolution.

B It generates new alleles that may be useful.

C It decreases genetic diversity within a population.

D It prevents natural mutations that may be harmful to the organism.

Question 9

The domestication of pigs (*Sus scrofa domestica*) from their wild ancestor, the wild boar (*Sus scrofa*), has occurred over approximately 10 000 years. During that time, humans have selected for and against particular traits.

The table below compares features of *Sus scrofa domestica* with those of *Sus scrofa*.

Trait	*Sus scrofa domestica*	*Sus scrofa*
Tusk length in males (cm)	8–15	12–45
Adult shoulder height (cm)	75–95	75–80
Adult mass (kg)	140–350	60–175
Litter size	10	10
Age of sexual maturity in females (months)	5–6	7–12

Based on the information above, humans have selected *against* large

A mass.

B height.

C litter size.

D tusk length.

Question 10

Cloning via somatic cell nuclear transfer (SCNT) has been successfully carried out in a variety of species, including many mammals. However, it is yet to be carried out successfully in birds. Scientists attribute this to two factors.

- The bird egg is much bigger than that of mammals, and the nucleus is located somewhere within the yolk.
- The bird embryo develops outside the mother's body, in a shelled egg.

Why is it difficult to clone a bird by SCNT?

A Birds do not have somatic cells.

B It is difficult to extract a nucleus from the bird that is to be cloned.

C The egg would not provide sufficient nourishment for the developing clone.

D There is no equivalent of a uterine wall in which to implant the developing embryo.

Section II: Short-answer questions

Instructions to students
- Answer all questions in the spaces provided.

Question 11 (4 marks)

Complete the table below to describe the processes and outcomes of artificial insemination and artificial pollination.

	Artificial insemination	Artificial pollination
Processes		
Outcome, including a specific example		

Question 12 (7 marks) ©NESA 2002 SII Q29a (ADAPTED)

The diagram illustrates three steps involved in the polymerase chain reaction (PCR).

Step 1

Step 2

5' 3'
3' 5'
3' 5'
5' 3'
5' 3'
3' 5'

Step 3

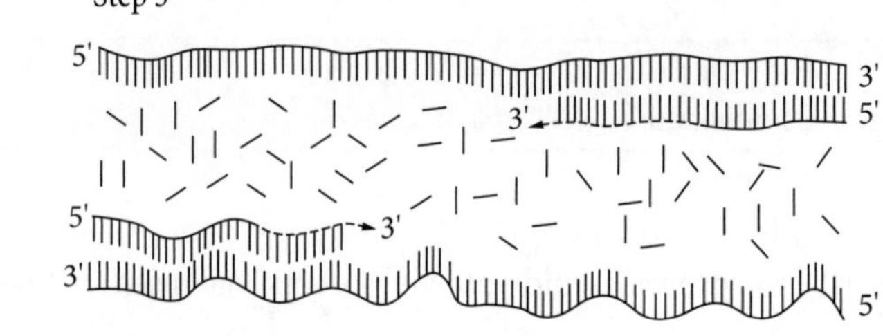

a Describe **one** use of PCR. 2 marks

b State what is happening in each of the steps shown in the diagram above. 3 marks

c In PCR, the three steps shown are repeated over and over again, in a cyclical fashion. Explain why the process is considered to be a 'chain reaction'. 2 marks

Question 13 (6 marks)

A farmer is looking to preserve the genetics of his prize-winning dairy cow. For the past eight years, the cow has birthed healthy calves, produced high-quality milk and remained in good health, despite others in the herd becoming ill. As she is reaching the end of her reproductive life, the farmer is considering the merits of cloning.

Discuss the famer's decision to use somatic cell nuclear transfer rather than embryo splitting.

Question 14 (4 marks)

Cystic fibrosis (CF) is an autosomal recessive disorder that causes severe damage to the lungs, digestive system and other organs in the body. Most men with CF will suffer from infertility, because the tube that carries sperm from the testicles to the penis is either missing or blocked.

In vitro fertilisation (IVF) is often used to help males with CF conceive. Embryos can be screened for CF prior to implantation, in a process called pre-implantation genetic diagnosis. If an embryo is found to have two affected CF alleles, it is discarded. The steps in this process are shown in the diagram below.

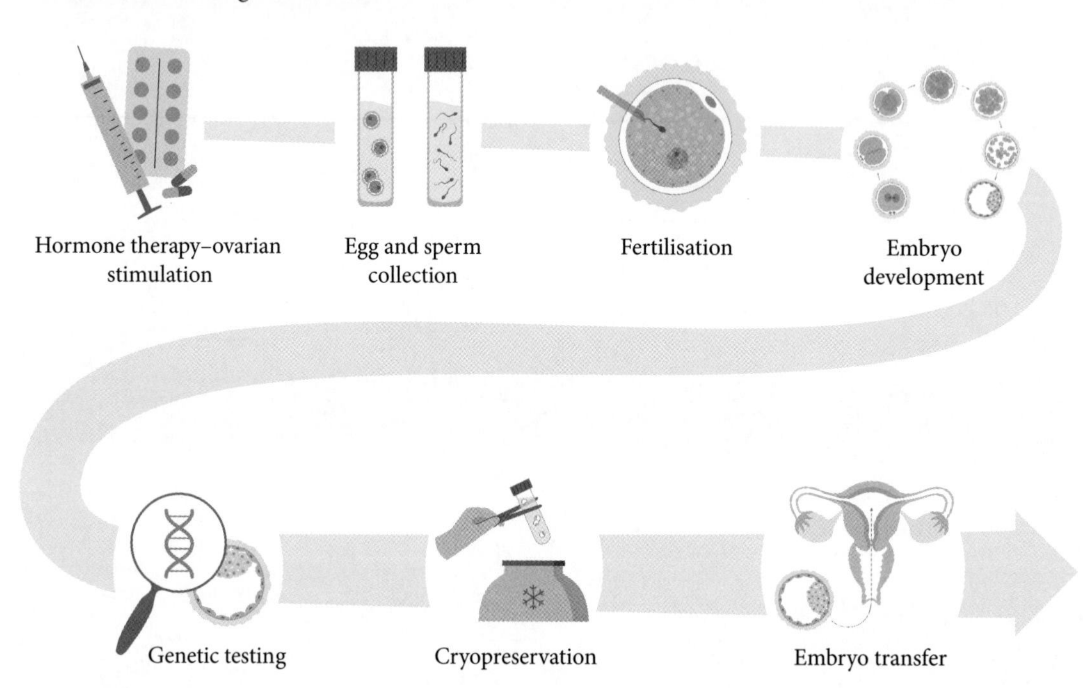

Evaluate **two** benefits of using IVF and pre-implantation genetic diagnosis in males with CF.

Question 15 (7 marks) ©NESA 2018 SII Q32c (ADAPTED)

The diagram below shows how a scientist inserted the gene for the production of human insulin into bacterial plasmids.

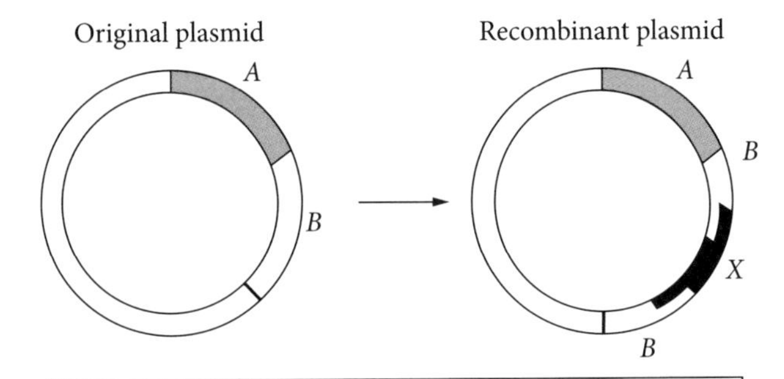

KEY:
Gene *A*: codes for resistance to the antibiotic ampicillin
Gene *B*: codes for resistance to the antibiotic tetracycline
Gene *X*: codes for the production of human insulin

a Outline the role of restriction enzymes in the production of recombinant plasmids. 2 marks

b Outline the role of DNA ligase in the production of recombinant plasmids. 1 mark

c Once the scientist created the recombinant plasmids, they incorporated them into host bacteria, in a process called 'transformation'. The host bacteria that received the plasmid normally have no resistance to either ampicillin or tetracycline antibiotics.

The scientist allowed the host bacteria to multiply in a nutrient broth overnight. They then streaked the broth onto several agar plates and incubated them to produce colonies that were clones of the host bacteria.

Explain how the scientist could use the antibiotics ampicillin and tetracycline to determine whether the recombination and transformation processes were successful. 4 marks

CHAPTER 3
MODULE 7: INFECTIOUS DISEASE

Test 9: Causes of infectious disease

Section I: 10 marks. Section II: 29 marks. Total marks: 39.
Suggested time: 70 minutes

Section I: Multiple-choice questions

Instructions to students

- For each question, circle the multiple-choice letter to indicate your answer.

Question 1

Which of the following is an example of a pathogen?

A Bacteria living within the intestinal tract, enabling digestion of some foods

B Fungi living in the urogenital tract, preventing growth of harmful bacteria

C Modified viruses in the body, targeting and killing cancer cells

D Protozoa ingested in water, causing diarrhoea and vomiting

Question 2 ©NESA 2021 SI Q16

Scientists conducted an experiment to investigate the effectiveness of treating water from storage dams with UV radiation.

The experiment was conducted more than three times. The results are shown in the table.

What conclusion may be drawn from the data obtained?

A The control plates are contaminated.

B High doses of UV eliminate all pathogens.

C Exposure to UV inhibits reproduction of bacteria.

D The presence of bacteria reduces the amount of UV.

UV dose (mJ/cm^2)	Agar plates after being inoculated with 5 mL of water and incubated at 25°C for 24 hours	
	Photo of agar plate	Average number of bacterial colonies counted
0		365
1		55
2		2

Question 3

What is the independent variable in the experiment in Question 2?

A UV dose

B Average number of bacterial colonies counted

C Incubation temperature

D Inoculation with 5 mL of water

Question 4

In July 2020, a team from the World Health Organization was sent to Kismayo, Somalia, after reports of 11 children with symptoms typical of a measles infection. Their investigation found a further 400 cases of measles in the community.

How could the measles outbreak in Kismayo be described?

A An epidemic

B A pandemic

C Endemic

D Eradicated

Question 5

Samples from a patient with meningitis were collected and examined under a microscope. They showed the presence of a single-celled prokaryotic pathogen that had a cell wall but no membrane-bound organelles and was approximately 5 μm in length. What type of pathogen was causing the patient's meningitis?

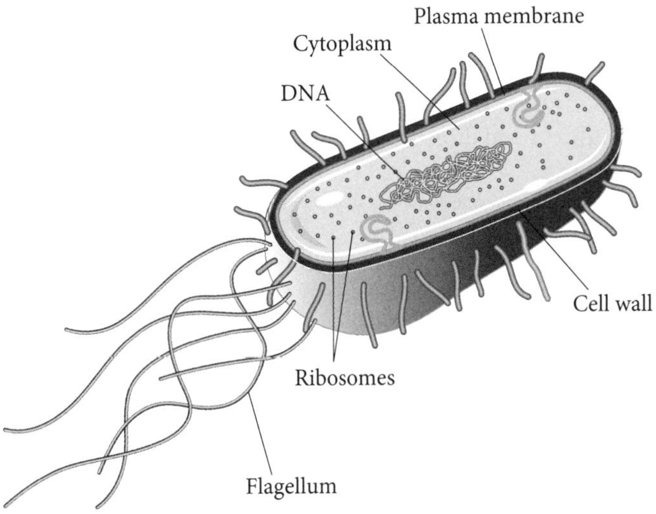

A Virus

B Prion

C Protozoan

D Bacterium

Question 6

Koch's postulates can be used to determine the cause of a disease. Syphilis is thought to be caused by the bacterium *Treponema pallidum*. Scientists have not been able to grow this bacterium in pure culture, but blood tests from sufferers indicate that it is present in every case of the disease.

Which of the following conclusions can be drawn about the cause of syphilis?

A Koch's postulates have confirmed that the bacterium *Treponema pallidum* is the cause of syphilis, because most of the criteria have been fulfilled.

B It is likely that syphilis is caused by the bacterium *Treponema pallidum*, but this has not been confirmed by Koch's postulates.

C The inability to grow the bacterium *Treponema pallidum* in pure culture means that it is definitely not the cause of syphilis, because Koch's postulates are not fulfilled.

D Koch's postulates should not be used to determine the cause of syphilis, because it is an infectious disease.

Question 7

Diseases that cause severe diarrhoea, such as cholera, are most commonly spread via the faecal–oral route, where a person unintentionally ingests faecal matter containing pathogens from an infected person.

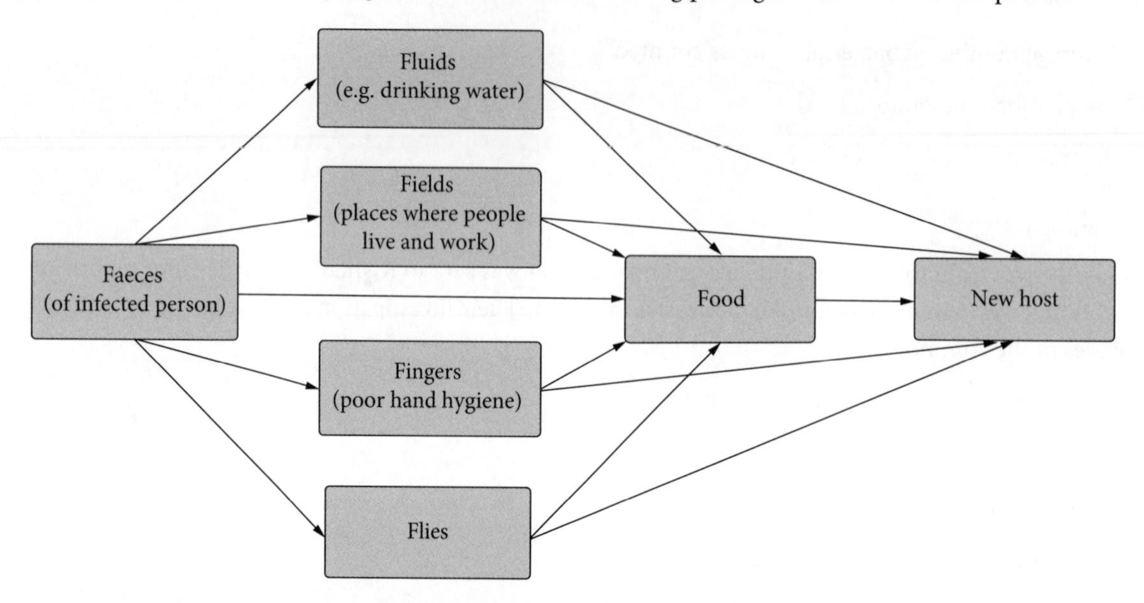

What mode or modes of transmission are shown in the diagram above?

A Indirect transmission

B Direct transmission

C Indirect and vector transmission

D Indirect and direct transmission

Question 8 ©NESA 2014 SI Q13

A student conducted a first-hand investigation using nutrient broth, beakers and an S-shaped delivery tube, in an attempt to model Pasteur's experiment.

The equipment and data collected are shown.

Test tube	Observation of nutrient broth Day 1	Observation of nutrient broth Day 14
1	Clear	Cloudy
2	Clear	Cloudy

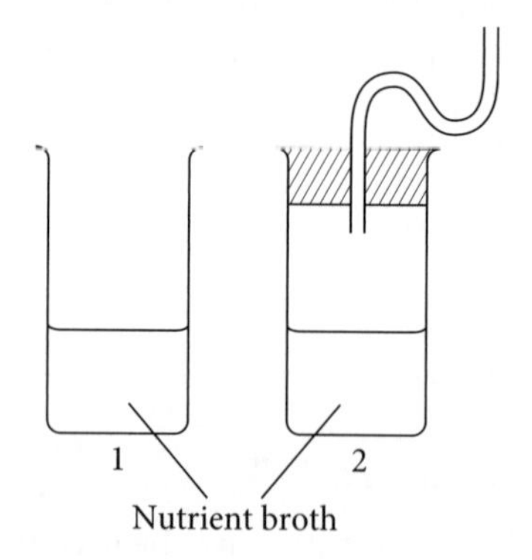

The student's results were different from Pasteur's results.

Which of the following provides the best explanation for the difference?

A The nutrient broth was different from Pasteur's.

B The nutrient broth always goes cloudy as it ages.

C The nutrient broth was not boiled thoroughly on Day 1.

D The nutrient broths were both exposed to oxygen from the outside air.

Question 9 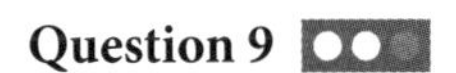

Which of the following is **not** an adaptation that facilitates a pathogen's transmission from one host to another?

A Ability to live for long periods of time in soil or water reservoirs

B Susceptibility to antimicrobial substances

C Formation of endospores to resist desiccation

D Ability to remain suspended in air for long periods

Question 10

Newcastle disease is a viral disease that infects birds, including chickens, causing loss of appetite, coughing, respiratory problems, nasal discharge and bright green diarrhoea. The disease causes reduced egg production and is highly transmissible between birds, often resulting in death. Newcastle disease is not endemic to Australia, but outbreaks have occurred in the past.

Why is Newcastle disease a concern to the agricultural industry in Australia?

A An outbreak would result in financial loss through reduced egg production and deaths of birds.

B Humans who eat meat from these chickens will contract Newcastle disease and die.

C The disease is not endemic, so farmers are unaware of the symptoms.

D It requires treatment with antibiotics, which is costly and may spread resistance.

Section II: Short-answer questions

Instructions to students
- Answer all questions in the spaces provided.

Question 11 (3 marks) 

Complete the table below to classify pathogens that cause disease in plants and animals.

Pathogen	Cellular or non-cellular	Description of pathogen	Disease caused by this type of pathogen
Protozoa			
Prion			
Virus			

Question 12 (5 marks)

Agriculture is an essential primary industry in Australia. Infectious plant diseases and pests can have catastrophic effects on this industry.

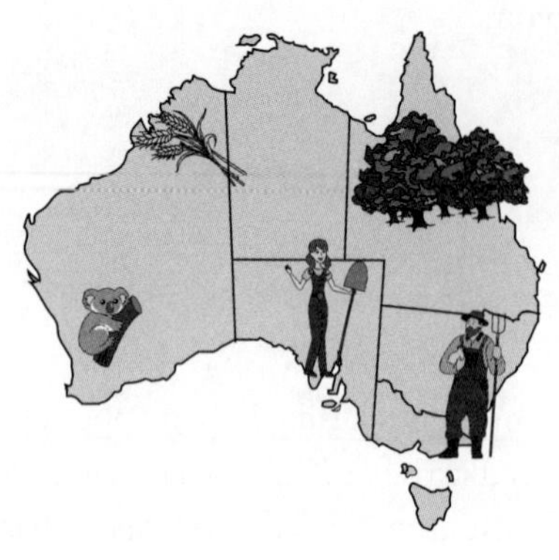

Risks to Australia from infectious plant diseases and pests:

- Risks to crops (broadacre and horticulture) ($29.3 billion)
- Risks to the livelihood of >63 000 people employed in horticulture industry
- Risks to forestry industry ($2.7 billion)
- Risks to Australia's unique environment

Using named examples, assess the causes and effects of plant diseases on agricultural production.

9780170465250

Question 13 (6 marks)

In 1859, Louis Pasteur conducted a now famous experiment that refuted the popular theory of spontaneous generation.

a With the use of diagrams, describe the experiment conducted by Pasteur, and discuss his contribution to our understanding of the cause and transmission of disease. 4 marks

b When conducting experiments using microbes, precautions must be taken to ensure safety. Construct a table to assess the risks that must be considered when conducting investigations using microbes. 2 marks

Question 14 (7 marks)

Zika virus commonly causes symptoms of fever, joint pain, muscle pain, headache and rash. It is rarely fatal. Zika is most often transmitted by the bite of the *Aedes* mosquito but can also be transmitted sexually or by blood transfusion. In 2015–2016, an outbreak of Zika virus occurred, with an estimated 1.5 million cases. The outbreak was first reported in Brazil and spread to a number of countries in South America, Central America and North America.

In 2016, the Olympic Games were held in Rio de Janeiro, Brazil. Rather than cancelling the games, as proposed by authorities, the decision was made to postpone the games until winter.

a Explain why health officials would be concerned about the Olympic Games being held in Rio de Janeiro during the outbreak of Zika virus. 2 marks

b Suggest why the games were held during winter. 2 marks

c The graph below shows the number of pregnant women with Zika virus in Brazil from 2015 to 2016, along with the number of babies born with microcephaly. Microcephaly is characterised by a small head and a small, underdeveloped brain.

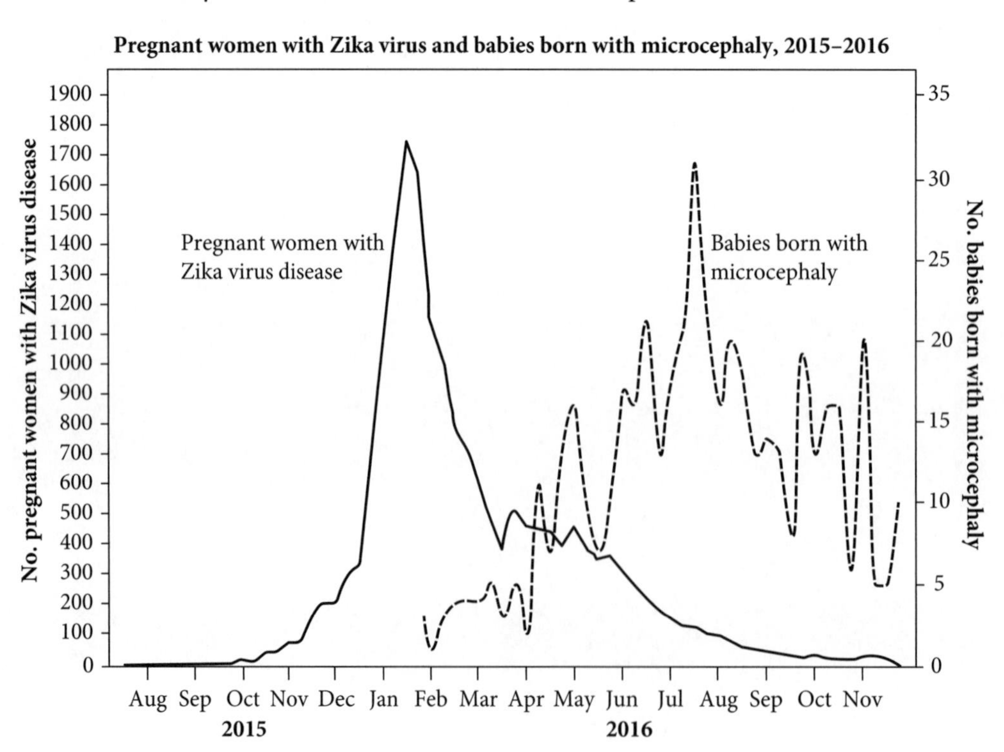

Describe the relationship between Zika infection during pregnancy and infant microcephaly shown in the graph and suggest what this means for the mode of transmission of the Zika virus. 3 marks

Question 15 (8 marks) ©NESA 2006 SII Q26

> Two Australian scientists, Robin Warren and Barry Marshall received Nobel Prize medals in 2005 for their discovery that the bacterium, *Heliobacter pylori*, is the main cause of stomach ulcers and gastritis.
>
> Previously, these conditions were thought to be caused by stress and bad eating habits.
>
> 'We were taught that bacteria don't grow in the stomach, but these did', said Warren of the bacterium, *Heliobacter pylori*.

The Weekend Australian Magazine, 10–11 December 2005

Warren and Marshall's investigations included:

- using a microscope to look at prepared slides of ulcerated stomach tissues;
- using a flexible endoscope to look into the stomachs of patients with stomach ulcers and gastritis (localised or general inflamation of the stomach);
- using staining techniques to determine the possible presence of bacteria in stomach tissue.

In addition, Warren checked that Marshall's stomach contained no *Heliobacter pylori*. Marshall then swallowed a dose of the bacteria, triggering symptoms of gastritis.

Assess the procedures that Warren and Marshall used to identify and confirm their conclusion about the pathogen that caused stomach ulcers and gastritis.

Test 10: Responses to pathogens

Section I: 10 marks. Section II: 27 marks. Total marks: 37.
Suggested time: 70 minutes

Section I: Multiple-choice questions

Instructions to students
- For each question, circle the multiple-choice letter to indicate your answer.

Question 1

Pathogens often release enzymes to break down a plant's cuticle. As a response, many plants have developed a thickened cuticle that the enzymes are unable to penetrate.

What type of defence is this?

A A passive, physical barrier

B A passive, chemical barrier

C A rapid active response

D A delayed active response

Question 2

Plants do not have an immune system in the same way that animals do. However, some plants can learn to recognise a pathogen that they have previously been infected with.

Which of the following is an example of this?

A Chemical receptors on the surface of a plant cell can detect the presence of a pathogen.

B Leaves hang vertically to avoid accumulating a water film that could act as a reservoir for pathogens.

C Salicylic acid acts as a signalling agent for subsequent infections, limiting the severity.

D Small stomata inhibit the entry of pathogens, only after a primary infection.

Question 3

Which of the following responses to pathogens occurs in both plants and animals?

A The cell wall acts as a physical barrier to prevent the entry of pathogens.

B Apoptosis (programmed cell death) results in a cluster of dead cells surrounding the pathogen and isolating it.

C Antibodies are produced, and these attach to pathogens and deactivate them.

D Histamine is released to initiate the inflammatory response.

Question 4 ©NESA 2009 SI Q3

Which of the following prevent entry of pathogens into the human body?

A The skin and phagocytosis

B The skin and chemical barriers

C Inflammation response and phagocytosis

D Inflammation response and chemical barriers

9780170465250

Question 5 ©NESA 2010 SI Q3

Cilia prevent the entry of pathogens into the human body by

A providing a protective body covering.

B producing secretions toxic to pathogens.

C moving trapped pathogens to the mouth.

D increasing circulation of blood to the infected area.

Question 6

What is the role of histamine in response to an infection?

A To inhibit the growth of pathogens, inactivate enzymes and increase body temperature

B To lower the pH of the blood in order to inhibit reproduction of pathogens

C To stop the bleeding from a wound and re-establish a barrier to prevent entry of pathogens

D To trigger the dilation of blood vessels and increase their permeability

Question 7

A student constructed the table below, to summarise the role of various cells in the immune response, but she made a number of errors in the table.

Which row of the table is completed correctly?

	Name of cell	Role	Location
A	Mast cell	Antigen-presenting cell Phagocytosis of pathogens and cancer cells	Circulating in blood
B	Macrophage	Blood vessel dilation Release of histamine	Migrates from blood vessels into tissue
C	Neutrophil	Releases chemicals to inhibit pathogens Recruits other cells to site of infection	Migrates from blood vessels into tissue
D	Monocyte	Defence against parasites Release of histamine Produces allergic reactions	Circulating in blood

Question 8

The normal body temperature for humans is around 37°C. In response to a pathogen, the body may react by raising the 'set point' in the hypothalamus and allowing the body temperature to be raised. This is known as a fever.

What is the name of the fever-causing chemicals that allow this to happen?

A Interleukins

B Cytokines

C Monocytes

D Pyrogens

Question 9

Which of the following lists **only** physical barriers to prevent the entry of a pathogen?

A Mucous membranes, tight junctions between cells, sphincter muscles in the urethra

B Oil on the skin, tears, mucus

C Peristalsis to move food through the oesophagus, vomiting, gastric secretions of the stomach

D Enzymes in saliva, fatty acids in sweat, oil on the skin

Question 10

A common symptom of a kidney infection is the more frequent release of small amounts of urine.

What type of response is this and what purpose does it serve?

A A chemical response because more urea is produced as a result of infection

B A chemical response because chemicals are released to kill bacteria

C A physical response because urination flushes pathogens out of the body

D A physical response because urine contains phagocytes to deactivate bacteria

Section II: Short-answer questions

Instructions to students

- Answer all questions in the spaces provided.

Question 11 (6 marks)

In response to the many viral and fungal pathogens that can infect them, many plants have developed adaptations to protect them from invasion.

a Outline the response of a named Australian plant to a named pathogen. 2 marks

b Design a practical investigation that could be used to investigate the response of the plant to the pathogen you named in part **a**. In your response, include a step-by-step method and a risk assessment. 4 marks

Question 12 (3 marks) ©NESA 2014 SII Q21 (ADAPTED)

a The diagram shows a process that is part of the immune response.

What is the name of the process? 1 mark

____

b Outline how inflammation contributes to the immune response. 1 mark

____

____

____

____

c Identify **one** type of cell involved in the processes above. 1 mark

____

Question 13 (4 marks) 

Bacterial pathogens such as *Escherichia coli* can be transmitted through contaminated food or water.

a Explain how the secretions of the stomach can protect the host against infection by these pathogens. 2 marks

____

____

____

____

b Suggest how **one** other physical response could be used to protect the host from an *E. coli* infection. 2 marks

____

____

____

____

Question 14 (6 marks)

Phytophthora cinnamomi is a fungal pathogen that infects and damages the roots of native Australian trees, preventing them from absorbing water and causing dehydration of the plant. Dehydration of the plant results in death, known as 'dieback'. Scientists conducted a study in which seedlings of six species of *Acacia* plants were inoculated with the *P. cinnamomi* fungus. The data from this study are shown below.

Acacia species	Proportion of plants that showed signs of dieback (%)
1	7
2	47
3	0
4	14
5	80
6	20

a Construct a graph of the information shown above. 3 marks

b In response to the *P. cinnamomi* pathogen, some species of *Acacia* release an enzyme that causes lignin production to increase, strengthening the cell walls of root cells.

Explain how this adaptation aids survival of the plant and identify which species above is most likely to have developed this adaptation. 3 marks

Question 15 (8 marks)

While a woman was removing a wooden fence from her garden, her finger was pierced by a large splinter.

Describe the role of skin in defending the body against pathogens, and account for all the physical and chemical responses that would occur in this woman's body to prevent the spread of infection.

Test 11: Immunity

Section I: 10 marks. Section II: 28 marks. Total marks: 38.
Suggested time: 70 minutes

Section I: Multiple-choice questions

Instructions to students

- For each question, circle the multiple-choice letter to indicate your answer.

Question 1

An antigen can be defined as a

A chemical substance that inhibits the reproduction of pathogens.

B molecule that triggers an immune response in the host.

C protein produced by the immune system in response to a specific pathogen.

D protein involved in the production of lymphocytes during an immune response.

Question 2 ©NESA 2017 SI Q11

A student was asked to complete a table showing whether T cells and B cells have particular characteristics.

Which row did the student complete correctly?

	Characteristic	T cell	B cell
A	Produces plasma cells	✓	✓
B	Produces antibodies that are released in body fluids	✓	✗
C	Cell surface receptor can recognise a specific antigen	✓	✓
D	Forms clones once stimulated	✗	✓

Question 3

Which of the following statements is correct?

A The innate response is specific to a pathogen and has 'memory' to improve the immune response to pathogens on future exposure.

B The adaptive response is specific to a pathogen and has 'memory' to improve the immune response to pathogens on future exposure.

C The innate response is non-specific to a pathogen but produces memory cells to improve the immune response to pathogens on future exposure.

D The adaptive response is non-specific to a pathogen but produces memory cells to improve the immune response to pathogens on future exposure.

Question 4

The immune system must be able to respond to a wide range of pathogens, both inside and outside the cells.

Which table below correctly identifies the location of the cell-mediated response to each type of pathogen?

A

	Inside cells	Outside cells
Protozoa	✓	✗
Bacteria	✗	✓
Virus	✓	✓
Fungi	✗	✓

B

	Inside cells	Outside cells
Protozoa	✓	✗
Bacteria	✓	✗
Virus	✓	✗
Fungi	✓	✗

C

	Inside cells	Outside cells
Protozoa	✗	✓
Bacteria	✗	✓
Virus	✗	✓
Fungi	✗	✓

D

	Inside cells	Outside cells
Protozoa	✓	✗
Bacteria	✓	✗
Virus	✓	✗
Fungi	✗	✓

Question 5

In addition to pathogens, the immune system can recognise cancerous cells and respond to deactivate them. One such response is shown in the diagram below.

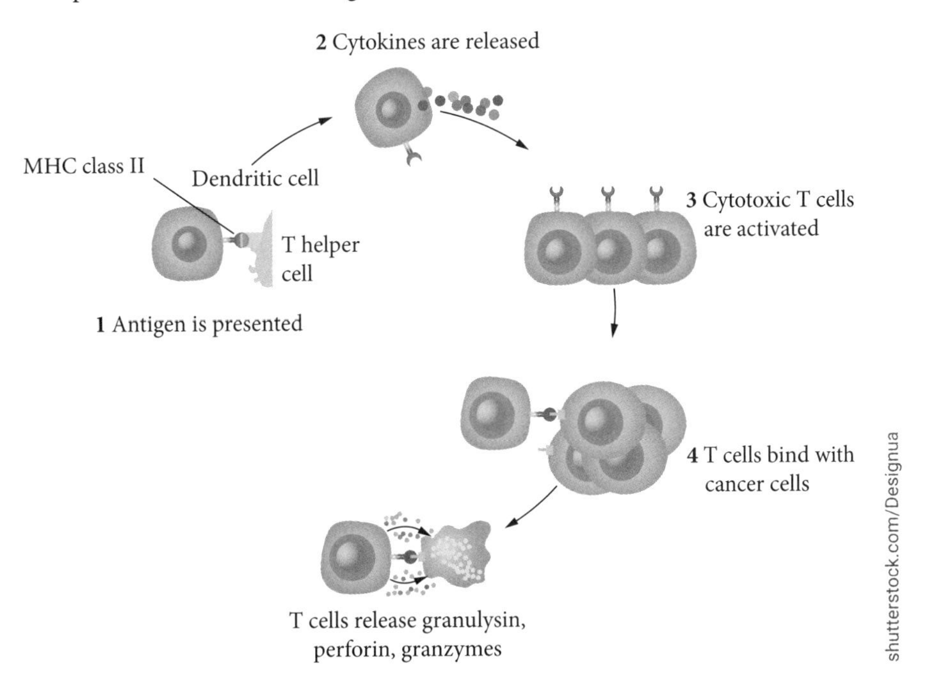

What process is occurring in the diagram, and is this part of the innate or adaptive immune response?

A This is programmed cell death and it is part of the adaptive immune response.

B This is programmed cell death and it is part of both the innate and adaptive immune responses.

C This is the humoral response to an antigen and it is part of the innate immune response.

D This is the humoral response to an antigen and it is part of both the innate and adaptive immune responses.

Question 6

An important feature of the adaptive immune response is the ability to destroy foreign molecules without reacting to the cells of the host.

Which feature of the adaptive immune system allows it to distinguish between 'self' and 'non-self'?

A Cytokines

B Interleukins

C MHC molecules

D Antibodies

Question 7

The diagram below shows the cell-mediated response to a pathogen.

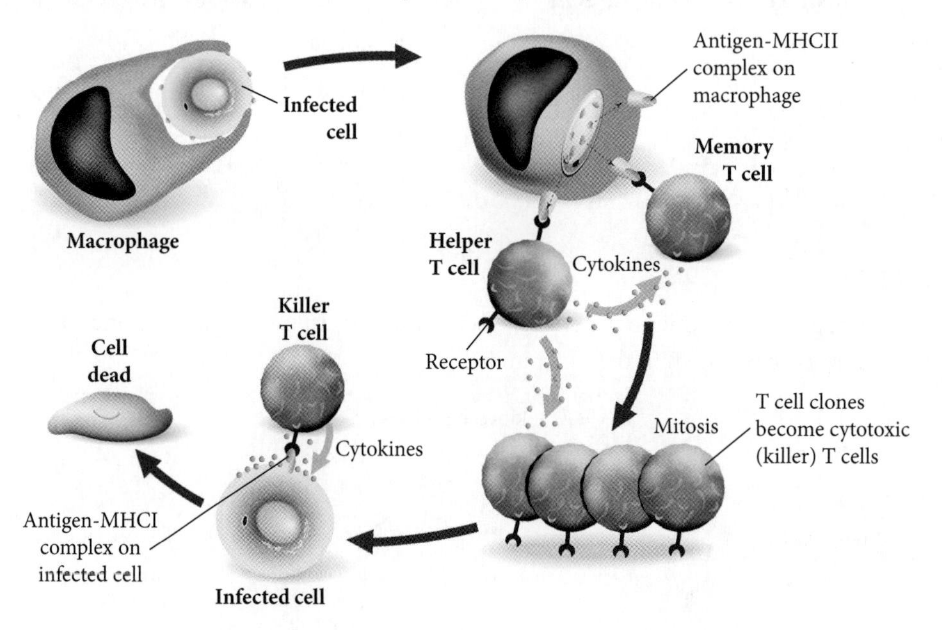

A website included the following statements related to the cell-mediated response.

I The cell-mediated response results in the production of antibodies.

II Helper T cells release cytokines to activate memory T cells.

III Macrophages present MHC markers of antigens on their surface.

IV Cytokines stimulate killer T cells, causing them to engulf the infected cell.

Which of the above statements contain correct information?

A I and III

B II and III

C I and IV

D II and IV

Question 8

Which of the following responses forms part of the adaptive response to an invading pathogen?

A Cytotoxic T cells migrate to the site of an infection, and their antigen receptors bind to the antigen displayed on their surface.

B Pyrogens are released, and this stimulates a fever, which kills pathogens.

C Histamines are released, causing dilation of the blood vessels, which brings more phagocytes to the site of infection.

D Lysozymes are secreted in sweat and saliva, breaking down the cell walls of invading bacteria.

Question 9

At birth, what type of immunity do babies have?

A Both the innate and adaptive immune responses are fully functional at birth.

B The innate immune response is fully functional, but the baby has no antibodies to any diseases.

C The innate immune system is fully functional, and the baby may have some temporary antibodies received from the mother through the placenta and breastfeeding.

D The baby's adaptive system is fully functioning if the mother has been vaccinated against all common diseases, but it takes time for the baby's innate immune system to develop.

Question 10 ©NESA 2015 SI Q17

The diagram shows an interaction between cells of the immune system.

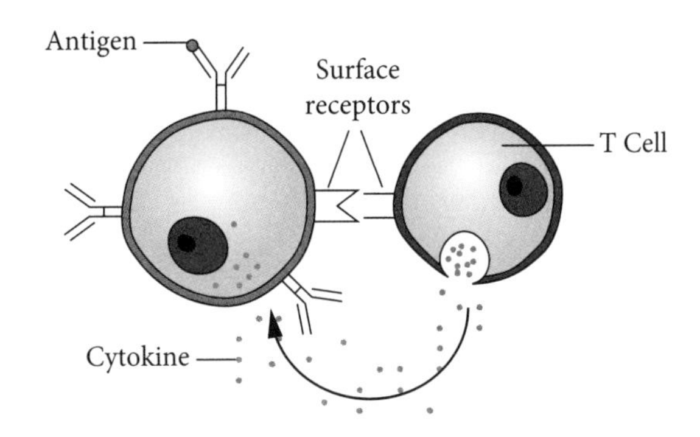

What specific process is shown in the diagram?

A B cell encountering an antigen

B Activation of a macrophage by a helper T cell

C Stimulation of a B cell to become a plasma cell

D Cytotoxic T cell destroying a virus-infected cell

9780170465250

Section II: Short-answer questions

Instructions to students
- Answer all questions in the spaces provided.

Question 11 (5 marks)

Construct a flow chart in the space below to show the steps involved in the production of antibodies in order to provide long-term immunity to a disease after exposure to a pathogen.

Question 12 (10 marks) ©NESA 2010 SII Q28 (ADAPTED)

Organ transplants may trigger an immune response which can lead to organ rejection. The diagram below represents a model of two heart cells, one from a transplant recipient and one from a donor.

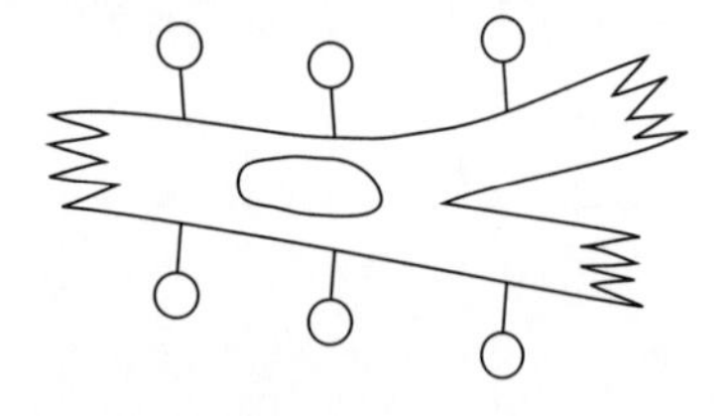

Recipient heart cell

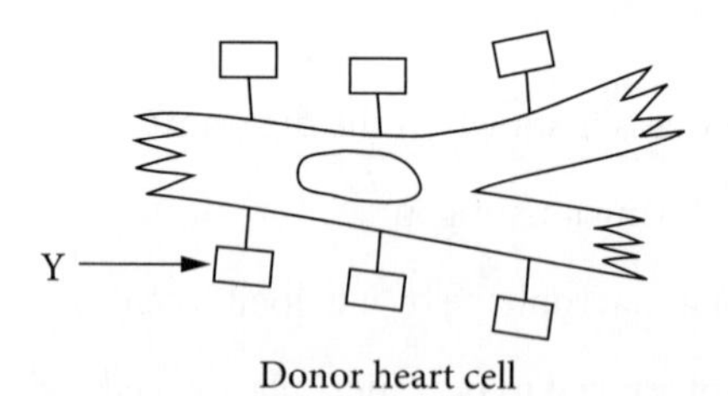

Donor heart cell

a What does Y represent? 1 mark

b Assess the effectiveness of the model in explaining the cause of organ rejection. 3 marks

c Why are models such as this used in Biology? 2 marks

d Name and outline the role of **two** types of T lymphocytes in organ rejection. 4 marks

Question 13 (5 marks)

a With the use of diagrams, describe the role of phagocytosis in responding to an invading pathogen. 3 marks

b Explain whether phagocytosis is part of the innate or adaptive immune response. 2 marks

Question 14 (4 marks)

a On the axes below, show the change in concentration of antibodies in the blood after a primary and a secondary exposure to a specific pathogen. 2 marks

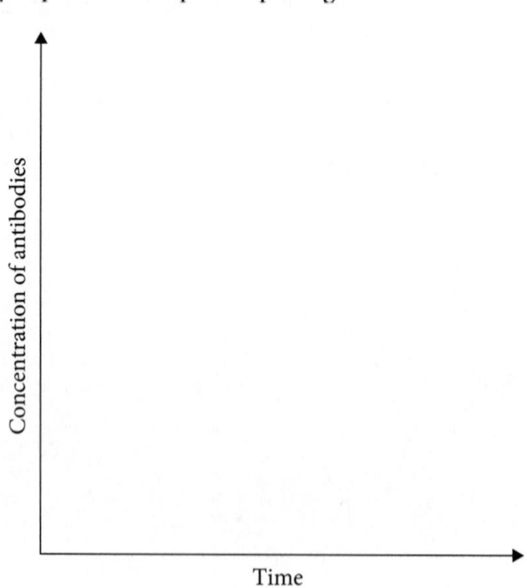

b Vaccinations provide immunity by activating the immune response through exposing an individual to the pathogen. Justify the administration of a second dose of most vaccinations commonly given in Australia. 2 marks

Question 15 (4 marks)

"The adaptive immune system provides more essential responses to a pathogen than the innate immune system."

Discuss the validity of this statement.

Test 12: Prevention, treatment and control

Section I: 10 marks, Section II: 28 marks. Total marks: 38.
Suggested time: 70 minutes

Section I: Multiple-choice questions

Instructions to students
- For each question, circle the multiple-choice letter to indicate your answer.

Question 1 ©NESA 2021 SI Q8

Howard Florey conducted a breakthrough experiment in the development and use of antibiotics. He infected eight mice with *Streptococcus* bacteria. Four mice were given penicillin and survived while the four untreated mice died.

What conclusion could be drawn from the data obtained?

A The experiment should be repeated with more mice.

B The use of penicillin causes antibiotic resistance in mice.

C Penicillin may be used on humans safely to treat bacterial infections.

D Penicillin may have played a role in the survival of the four treated mice.

Question 2

Foot and mouth disease (FMD) is a highly contagious viral disease that affects animals, including cattle, sheep and pigs. It causes fever as well as blisters on the feet, mammary glands and mouth. It can be lethal, especially in young animals. FMD is common in the Middle East, South-East Asia, Africa and South America, but Australia is free of FMD, as well as many other agricultural diseases, because of its isolation from other countries. To protect the agricultural industry from diseases such as FMD, it is important to take steps to prevent entry of the disease into Australia.

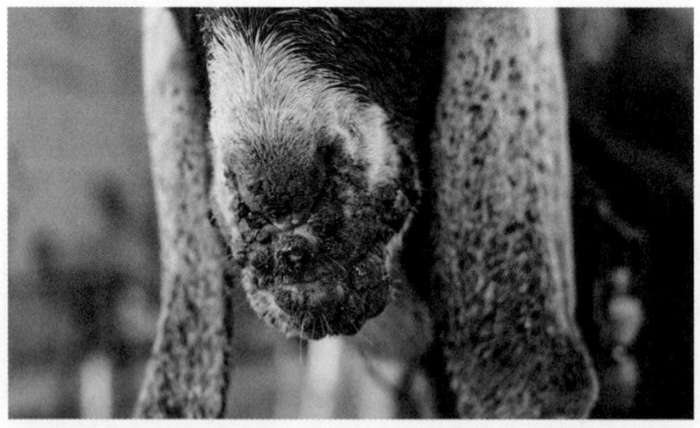

shutterstock.com/thala bhula

What method would be most effective in preventing the entry of FMD into Australia?

A Treat all animals on Australian farms with antiviral medication as a precaution, in case the disease is introduced to a farm.

B Train farmers in countries with FMD to recognise the symptoms and provide treatment with antibiotics before exporting animals or meat to Australia.

C Ensure that anyone involved in importing these animals into Australia wears a mask and gloves and washes their hands after touching the animals.

D Ban the import of live animals, uncooked meat and unprocessed dairy into Australia from countries with FMD.

Question 3

In 2020, the following graphic was released as part of a public education campaign in the UK to help prevent the spread of COVID-19.

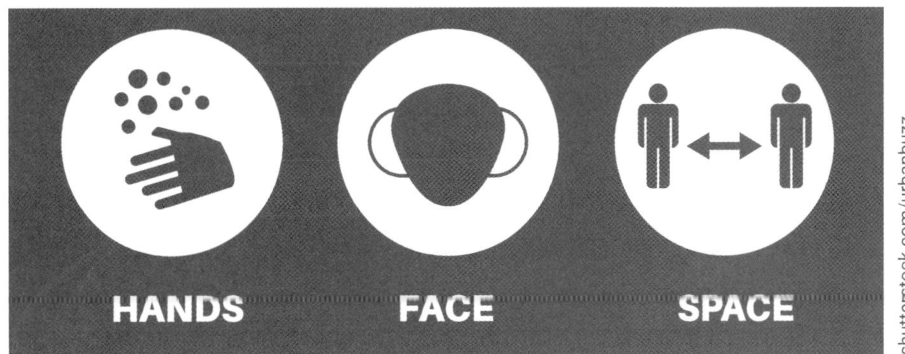

shutterstock.com/urbanbuzz

What aspects of this education campaign would make it effective?

A Following these three steps will definitely prevent a person from getting the virus.

B Following these steps will not prevent a person getting virus but may prevent hospitalisation.

C There is a detailed set of instructions for people to follow precisely.

D The message is clear, simple and easy to remember, so it is more likely to be followed.

Question 4

In 1918, the Spanish influenza pandemic began, ultimately killing more than 50 million people worldwide.

Despite considerable efforts to avoid entry of Spanish influenza into Australia, cases began to appear in 1919. However, strategies remained in place to minimise the number of cases and prevent the spread of the disease. One strategy was the interception of ships entering Australian waters by quarantine services, who checked all on board for signs of infection. Over the course of the pandemic, 323 ships were checked and 174 were found to have infected people on board. A total of 81 510 people were checked for infection, with 1102 found to be infected with the disease.

National Library of Australia N 614.518 OLI

What was the prevalence of Spanish influenza among those checked by quarantine services?

A 1.35%

B 5.39%

C 6.33%

D 8.15%

Question 5

In addition to monitoring symptoms among those entering by sea, authorities also closed schools and entertainment venues, and mandated the wearing of face masks to prevent the spread of influenza.

What effect would such measures have on the spread of the disease?

A These measures would be completely effective at preventing transmission of the disease between individuals.

B These measures would slow the spread of the disease, by minimising contact between individuals and preventing transmission via airborne droplets.

C These measures would have been unnecessary because a vaccine was developed that prevented death due to secondary bacterial infections.

D These measures would have been unnecessary because Spanish influenza was not easily spread between individuals.

Question 6

Imported plant products can potentially carry diseases and plant pests, which could have devastating effects on Australia's agricultural industry. Diseases can cause discolouration of flowers, spots on leaves, cankers and rotting of fruit before it ripens.

shutterstock.com/Dan Gabriel Atanasie

The image on the right shows a leaf with a viral mosaic disease. Which of the following methods would **not** protect Australian agricultural plants from imported viral plant diseases?

A Genetic engineering of plants with naturally disease-resistant genes using CRISPR Cas-9 technology

B Spraying crops with pesticides to kill pests

C Preventing the importation of any live plant material by checking at airports, seaports and mail services

D Burning virally infected plants that have been seized, to prevent entry into the community

Question 7

Monitoring for disease can occur at the local, regional and global levels. Which of the following is an example of a regional factor that can influence the spread of a disease?

A Increased rainfall leading to a malaria outbreak in a number of equatorial countries

B Transmission of a disease from farm animals to humans in a town

C Mobility of refugees who have received inadequate health care prior to migration

D Screening for tuberculosis by giving prospective migrants chest X-rays before they migrate to a new country

Question 8 ©NESA 2019 SI Q10

A group of islands are separated from each other by large stretches of water. Each island has its own policy on quarantine.

A nearby country is experiencing an outbreak of an infectious disease in its cattle.

An investigation is to be designed to find which of the quarantine policies operating on the islands is the most effective.

Which of the following would be a suitable design feature of the investigation?

A The control is the smallest island.

B The control is the infected number of cattle.

C The independent variable is the quarantine policy.

D The independent variable is the number of infected cattle.

Question 9

Lassa fever is an acute haemorrhagic fever that occurs in West African countries. Rodents are vectors for the virus. In Nigeria, spiritual healers use rodents for the preparation of charms and curses, and it is believed that this cultural practice of directly handling rodents has increased the transmission of the disease. Which of the following methods would be most effective at controlling the spread of the disease, in light of these cultural practices?

A Suggesting that residents set traps to capture the rodents

B Educating the public so that they understand the risks involved in handling the rodents and abandon harmful practices

C Genetically engineering the rodents so that they are no longer vectors for the virus

D Suggesting that spiritual healers use improved hygiene methods when preparing curses or charms

Question 10

Malaria is a disease that is caused by a protozoan and transmitted by female *Anopheles* mosquitoes. The rate of malaria has decreased in heavily affected countries, such as those in South-East Asia.

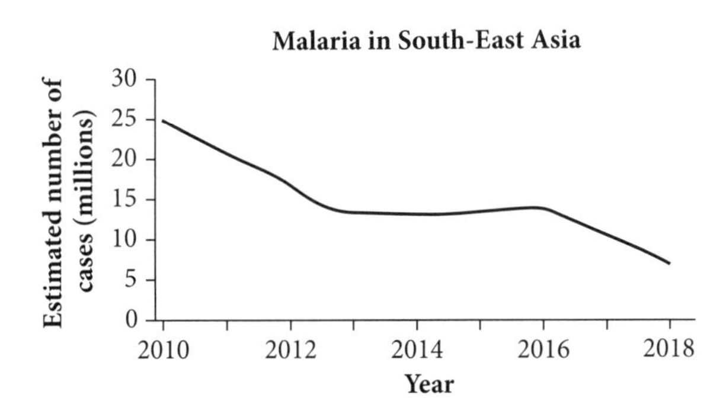

Data source: World Health Organization

What is a possible reason for the decrease in malaria cases in South-East Asia?

A Climate change has increased the temperature and precipitation, providing suitable breeding conditions for the mosquito vectors.

B Antiviral medications that can prevent infection with malaria have been developed.

C There has been widespread distribution of mosquito nets and effective public education campaigns.

D People infected with malaria have been quarantined until they are no longer infectious.

Section II: Short-answer questions

Instructions to students
- Answer all questions in the spaces provided.

Question 11 (6 marks)

In March 2020, nine cruise ships carrying more than 8000 passengers and crew were stranded at sea after they were denied entry into numerous seaports around the world. The ships were turned away due to fear that they were carrying the COVID-19 virus, which had just been recognised as a pandemic. Passengers and crew were forced to remain on board until they were given approval to dock.

a Describe steps that the crew of these cruise ships could have taken to prevent the spread of the virus on board the ships while they were stranded at sea. 3 marks

b Earlier that month, 2700 passengers and crew were allowed to disembark the *Ruby Princess* cruise ship in Sydney. Of the 2700 passengers, only 13 were tested as they left the ship, with 3 later testing positive. Untested passengers were allowed to travel home to other states and countries immediately. By the end of March 2020, a total of 440 passengers had tested positive to the virus, with 211 cases in New South Wales, 71 in South Australia, 70 in Queensland, 43 in Western Australia, 22 in the Australian Capital Territory, 18 in Victoria and 2 in the Northern Territory. Five deaths of passengers were also reported.

Outline how the mobility of the passengers on board the *Ruby Princess* may have affected the spread of the virus in Australia, and suggest precautions that could have been taken to prevent this. 3 marks

Question 12 (4 marks)

Indigenous Australians have long known about the benefits of biological materials for medicinal uses. This knowledge has become more widespread in recent years, with many pharmaceutical companies wanting to purchase the rights to use specific native plants to manufacture and sell pharmaceutical products to the general public. Using a named example, describe the use of **one** native plant for medicinal purposes, and discuss the importance of recognising and protecting Indigenous cultural and intellectual property.

Question 13 (5 marks)

In 1981, the first case of HIV/AIDS was reported; in 1983, scientists identified the human immunodeficiency virus as the causative pathogen. With no treatment available, there were a large number of deaths, and the disease was described as an epidemic. The virus hijacks CD4 cells, which are responsible for coordinating the immune response against other pathogens.

Medication was available from 1987 to slow the progression of the disease, but death was still inevitable. In 1996, anti-retroviral therapy was developed; this can suppress the viral load, allowing CD4 cells to function normally. Anti-retroviral treatment was available in all countries of the world by 2004, but there are continuing issues of inequity in accessing the treatment in some countries.

The data below shows the annual number of cases, infections and deaths from HIV/AIDS from 1990 to 2019.

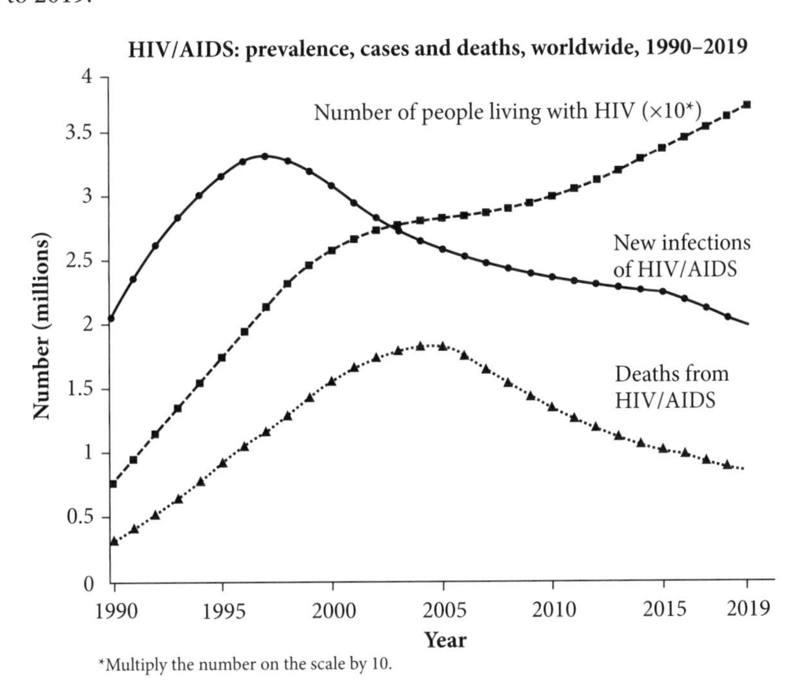

Data fom https://ourworldindata.org/grapher/deaths-and-new-cases-of-hiv

Assess the effectiveness of anti-retroviral therapy for the treatment and prevention of HIV/AIDS.

Question 14 (6 marks)

Dengue fever is spread by mosquitoes in the genus *Aedes*. While dengue fever is not currently endemic in Australia, mosquitoes in the *Aedes* genus live in warmer parts of Australia, and it is predicted that these mosquitoes will spread to other parts of Australia as climate change occurs. Experts warn that this could lead to dengue becoming endemic in Australia in the future. *Aedes* mosquitoes also spread other diseases, such as Ross River virus, which is endemic to Australia.

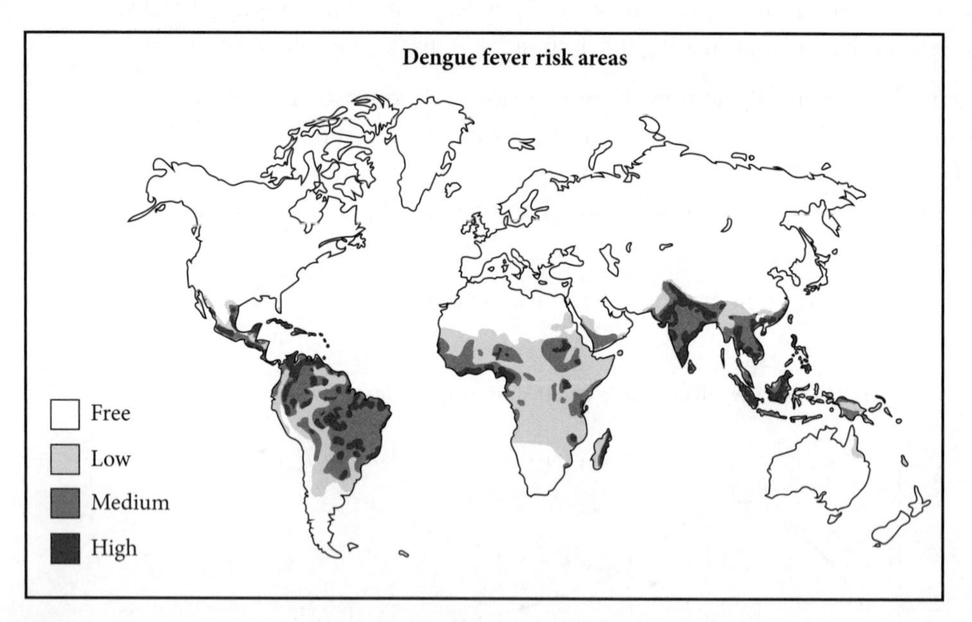

shutterstock.com/Alex Mit

a Explain how environmental management strategies could be used to control the spread of dengue virus and Ross River virus. 3 marks

b ●●● Ross River fever is currently the most widespread mosquito-borne disease in Australia. An effective vaccine has been developed, but authorities have determined that the benefits of the vaccine are not worth the financial cost to manufacture it.

Discuss whether this decision should be reconsidered in light of predictions about the spread of the *Aedes* mosquito to other parts of Australia. 3 marks

Question 15 (7 marks) ©NESA 2021 SII Q30 ●●●

A study compared the incidence of disease and survival of 8134 children who had received the measles vaccine with 8134 children from a neighbouring area who were unvaccinated against measles. Children in each group were matched for age, sex, size of dwelling, number of siblings and maternal education. The graphs show the number of measles cases among the two groups over three years.

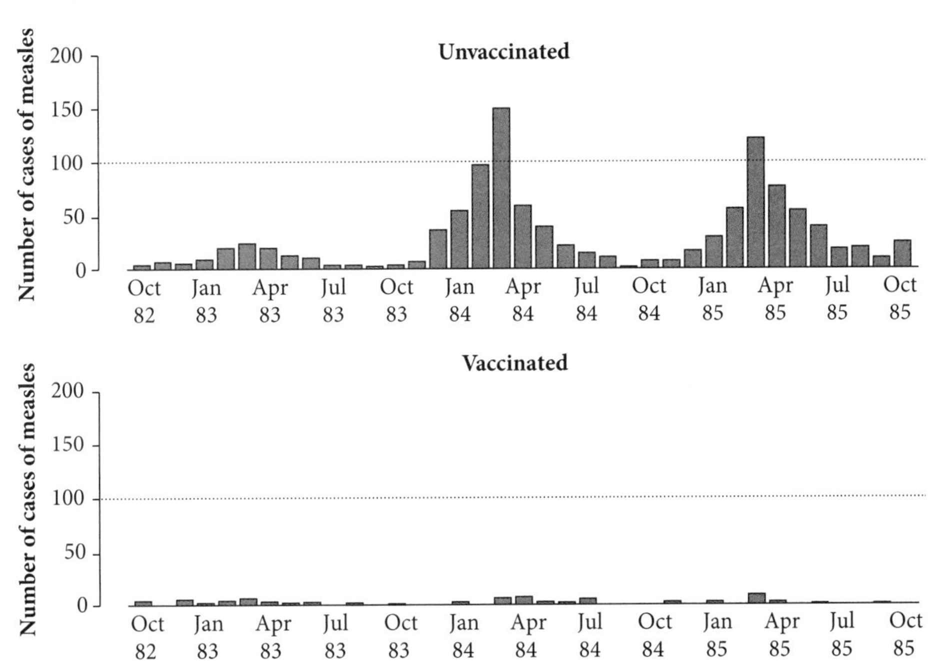

The table compares the cause of death and number of deaths of the two groups over the same three years.

Cause of death	Number of deaths	
	Children vaccinated against measles	Children unvaccinated against measles
Measles	2	40
Diarrhoea and dysentery	85	156
Oedema (swelling due to fluid in tissues)	6	21
Fever	22	25
Total	**115**	**242**

'A vaccine only protects the community against a specific disease.'

Analyse the data with reference to this statement.

CHAPTER 4
MODULE 8: NON-INFECTIOUS DISEASE AND DISORDERS

Test 13: Homeostasis

Section I: 10 marks. Section II: 26 marks. Total marks: 36.
Suggested time: 65 minutes

Section I: Multiple-choice questions

Instructions to students
- For each question, circle the multiple-choice letter to indicate your answer.

Question 1

Which of the following is an example of a physiological adaptation?

A Fur seals are insulated by a layer of fur.

B Snakes bask in the sun to absorb more heat.

C Bilbies burrow to avoid the heat of the day.

D Horses sweat to lose heat by evaporative cooling.

Question 2

Animals often have adaptations to help them lose or retain heat, to maintain their body temperature. The long-eared jerboa shown here has large, thin ears compared with its body size.

shutterstock.com/Lauren Suryanata

Which of the following statements correctly identifies the relationship between the ears of the long-eared jerboa and the probable environment it lives in?

A The ears have a large surface area to volume ratio, which would increase heat loss, so it probably lives in a hot climate.

B The ears have a large surface area to volume ratio, which would reduce heat loss, so it probably lives in a cold climate.

C The ears have a small surface area to volume ratio, which would increase heat loss, so it probably lives in a hot climate.

D The ears have a small surface area to volume ratio, which would reduce heat loss, so it probably lives in a cold climate.

Question 3 ©NESA 2021 SI Q13

The photographs show an open and a closed stomate on a leaf surface. When open, stomates allow water vapour to pass out of the leaf.

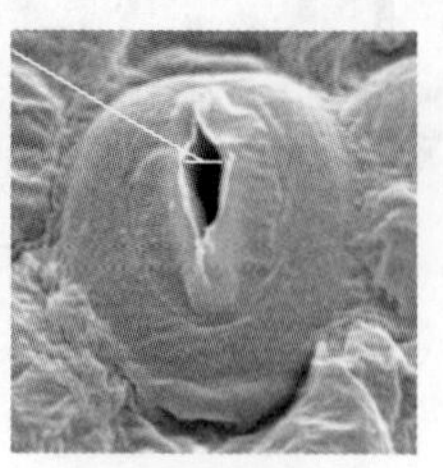

Open stomate

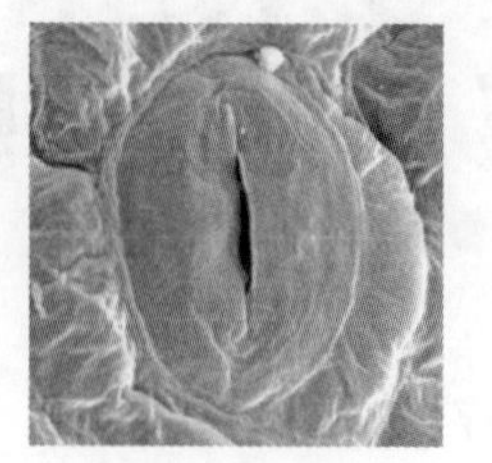
Closed stomate

Regulating stomates is a mechanism by which plants maintain water balance.

Which of the following graphs best illustrates this homeostatic mechanism?

A

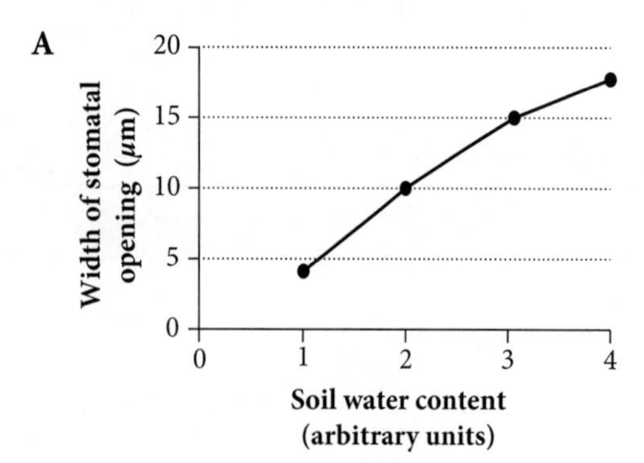

B

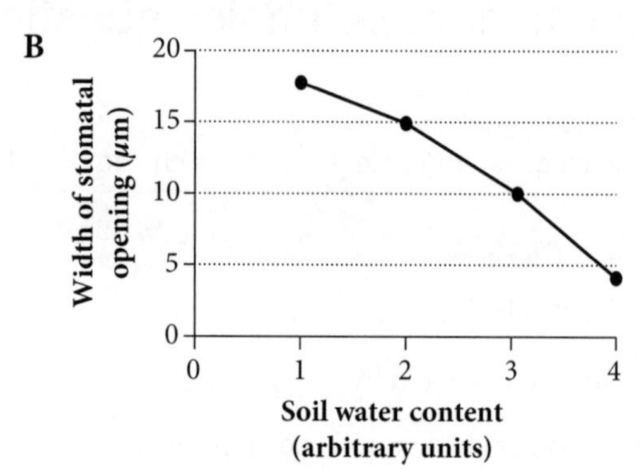

C

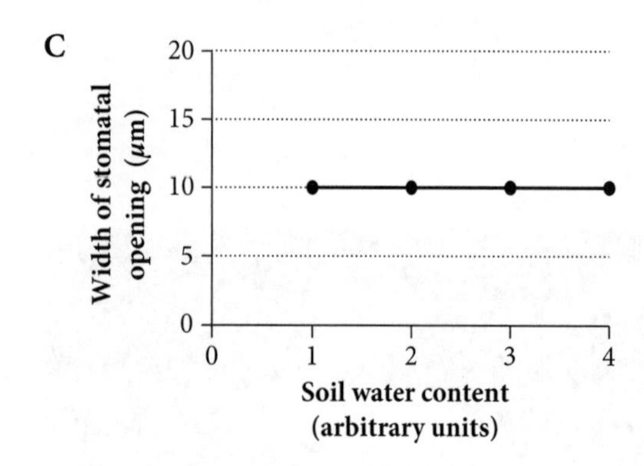

D

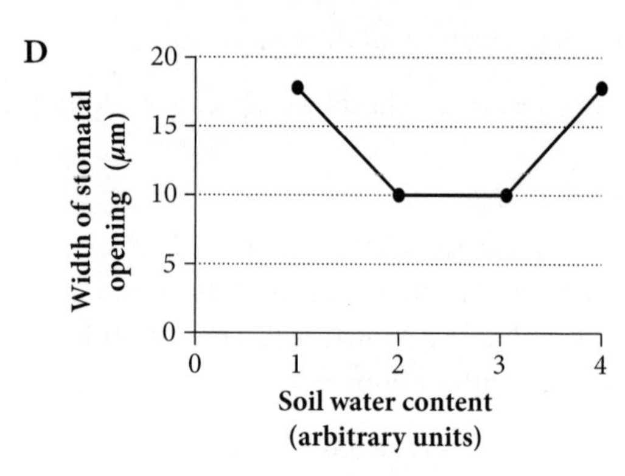

Question 4

Which of the following is an example of negative feedback?

A During childbirth, oxytocin is released at levels above the set point, to stimulate contractions of the uterine muscles.

B When someone steps on a sharp object, nerves send a message to the brain, resulting in the person feeling pain.

C In the presence of an infection, the body's temperature rises above the set point in order to create an inhospitable environment for the pathogen.

D When salt levels in the body are too high, aldosterone is released to bring the levels back to the set point.

Question 5 ©NESA 2015 SI Q7

The diagram shows a homeostatic mechanism in a mammal.

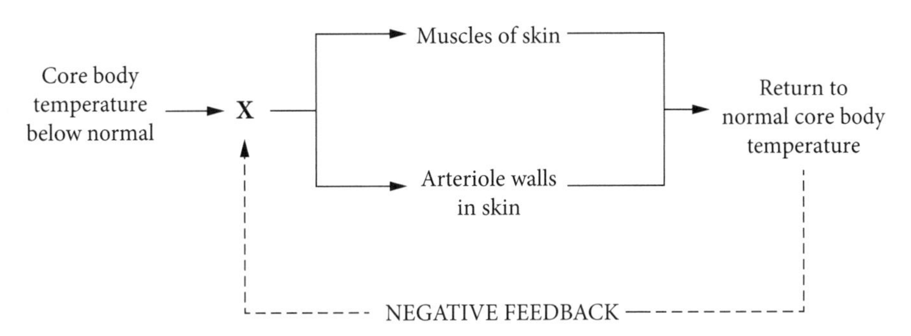

What does X represent in the diagram?

A The heart

B The brain

C A thermoreceptor in the skin

D A pressure receptor in a blood vessel

Question 6

Which system is responsible for secretion of the hormones responsible for maintaining homeostasis?

A Lymphatic system

B Endocrine system

C Integumentary system

D Nervous system

Use the information below to answer Questions 7–9.

Blood glucose levels need to be maintained at approximately 90 mg/100 mL in order to adequately supply the cells with glucose for respiration. The following negative feedback loop outlines the responses to changes in blood glucose levels.

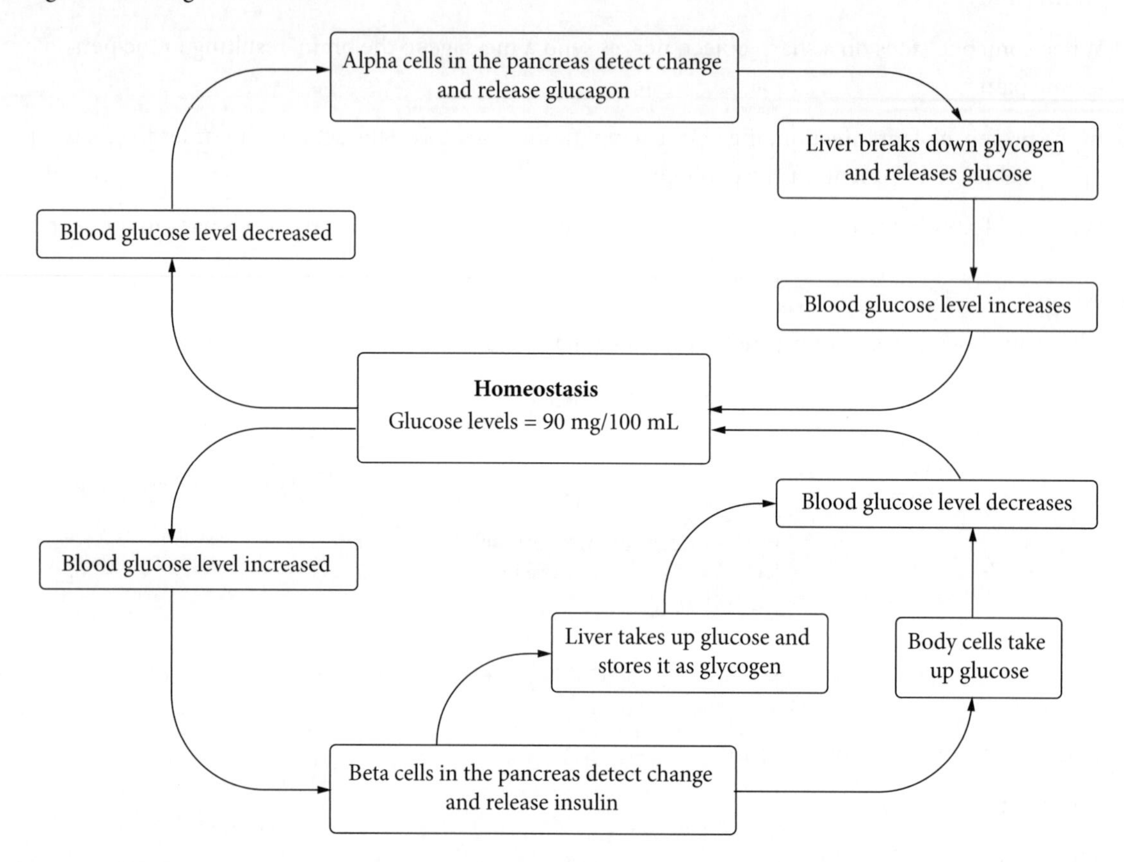

Question 7

Which row of the table correctly identifies the components of the negative feedback mechanism after a sugary snack is consumed?

	Receptor	Stimulus	Effector	Response
A	Liver cells	Blood glucose decreases	Alpha cells	Glucose levels increase
B	Alpha cells	Glucose is stored as glycogen	Beta cells	Release insulin
C	Beta cells	Blood glucose increases	Body cells	Take up glucose
D	Hypothalamus	Glycogen is broken down to glucose	Pancreas cells	Release glucagon

Question 8

Which hormone reduces the blood sugar level when it has decreased below the set point of 90 mg/100 mL?

A Insulin

B Glucose

C Glycogen

D Glucagon

Question 9

A person with diabetes does not produce sufficient insulin.

In the graphs below, a non-diabetic's blood sugar level is shown with a solid line, and a diabetic's blood sugar level is shown with a dotted line.

Which of the graphs correctly shows what would happen to a diabetic person's blood sugar levels after consuming a sugary snack, in comparison with a person without diabetes?

A

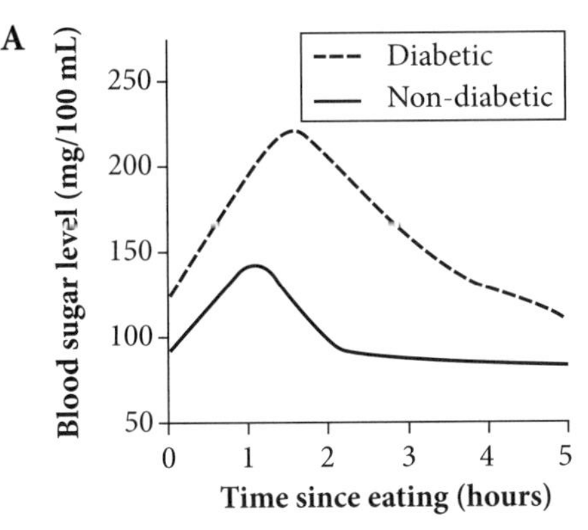

B

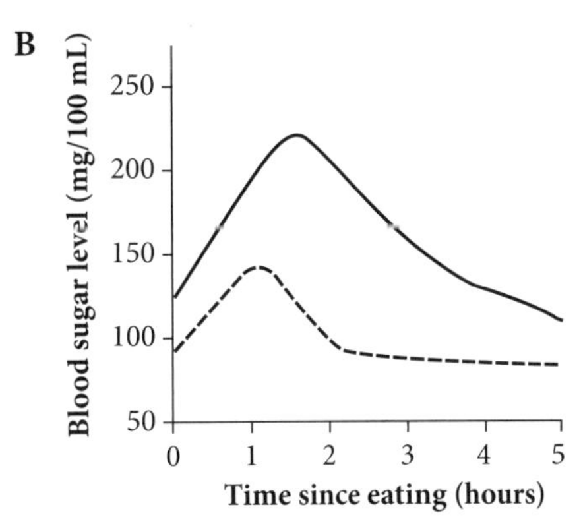

C

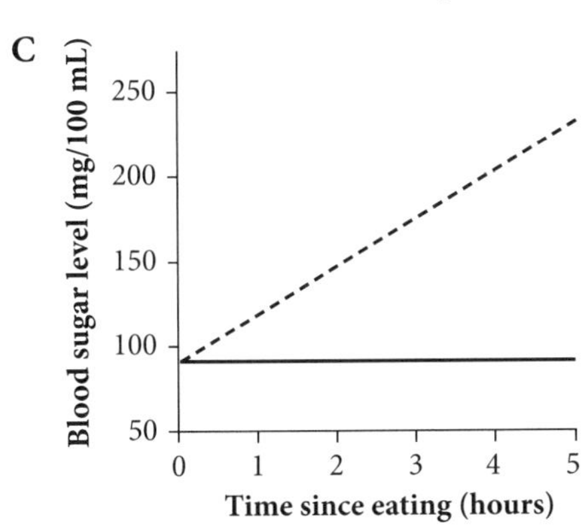

D

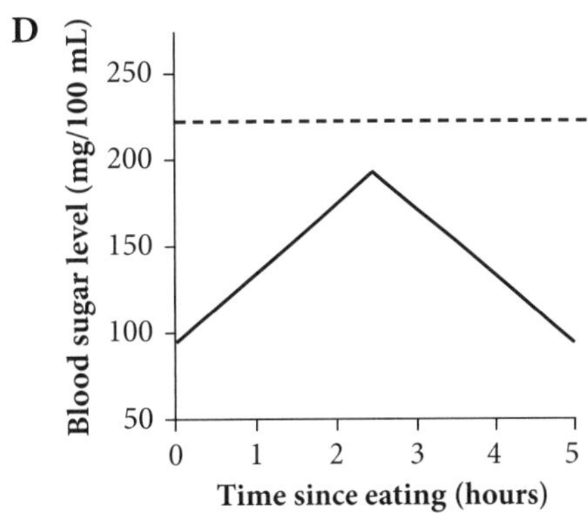

Question 10

Messages that allow for the maintenance of homeostasis travel by neural pathways provided by the nervous system, through the neurons.

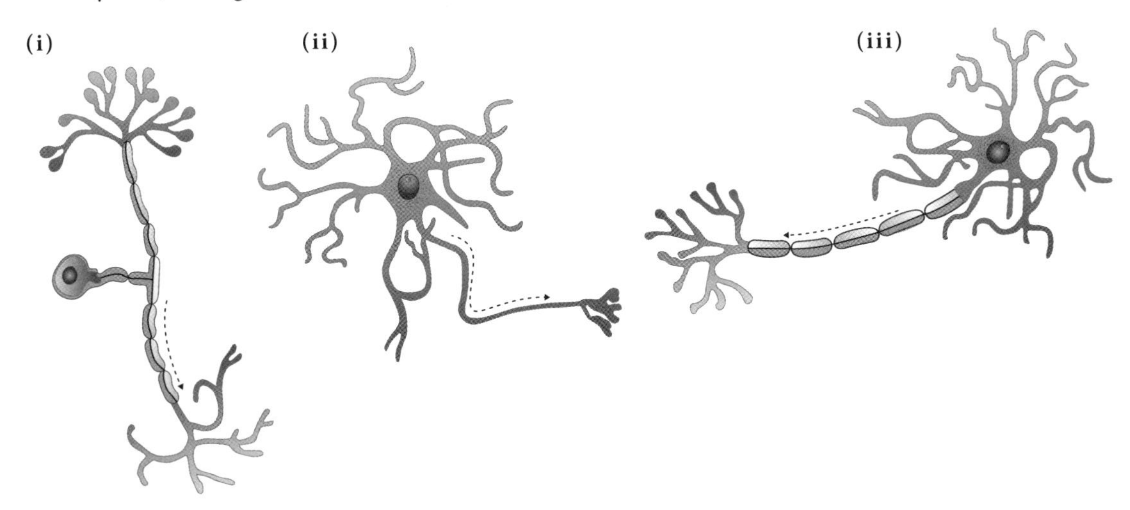

Which row in the table below correctly identifies the three types of neurons shown here?

	(i)	(ii)	(iii)
A	Sensory neuron	Motor neuron	Interneuron
B	Motor neuron	Interneuron	Sensory neuron
C	Sensory neuron	Interneuron	Motor neuron
D	Interneuron	Sensory neuron	Motor neuron

Section II: Short-answer questions

Instructions to students

- Answer all questions in the spaces provided.

Question 11 (6 marks)

Endotherms living in water may struggle to maintain their body temperature because water can conduct heat away from the body almost 25 times faster than air. Mammals that live in the ocean, including whales, dolphins, seals and sea lions, have developed an insulating layer, called blubber, that prevents loss of heat to the surrounding water.

Some scientists observed that the thickness of the blubber varies significantly and wondered whether the thickness is related to the temperature of the water. The data below was collected from sea lions in water of varying temperatures around the world.

Average water temperature (°C)	Average thickness of blubber (cm)
5	9.6
10	8.4
15	6.5
20	4.2
25	1.7

a On the grid below, plot a graph of the data in the table. 3 marks

b What conclusion should the scientists draw from the data collected? 2 marks

c What type of adaptation is the blubber layer of sea lions? 1 mark

Question 12 (5 marks)

a Outline the role of the nervous system in maintaining homeostasis. 2 marks

b The graph below shows changes in the electrical potential of the cell membrane for two stimuli, A and B.

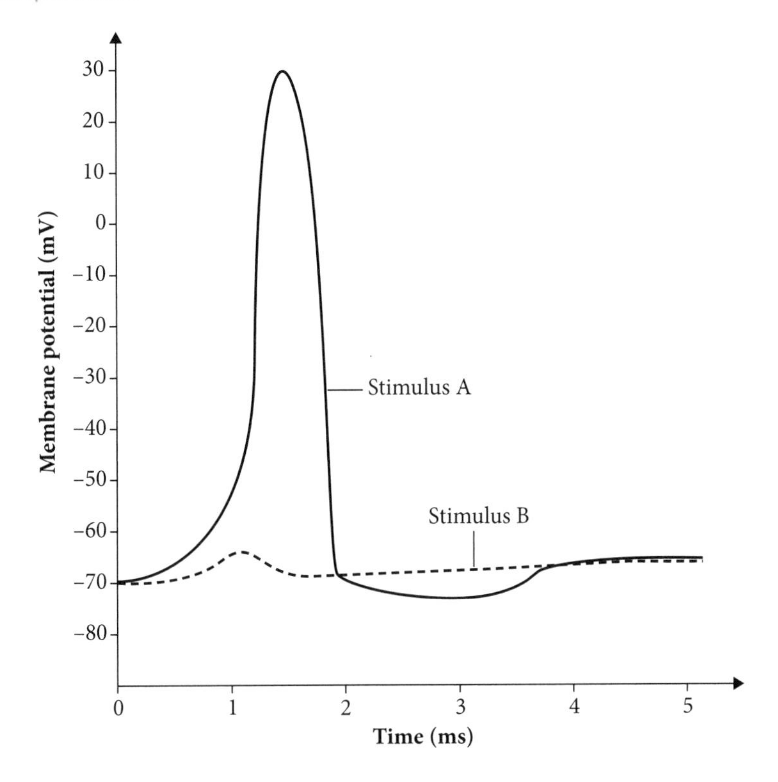

Would both stimuli A and B result in an action potential? Explain your answer. 3 marks

Question 13 (6 marks)

A marathon runner competed in a race on a summer day with a temperature of 32°C. When she completed the race, she immediately immersed herself in an ice bath to reduce muscle soreness. The temperature of the ice bath was 10°C.

Construct a negative feedback loop to show the changes that would occur in the marathon runner's body in response to the changes in temperature experienced during the race and the ice bath.

Question 14 (4 marks)

Many Australian plants have developed adaptations that allow them to maintain adequate water levels despite the hot and dry climate they grow in.

Explain how **two** mechanisms in plants allow for water balance to be maintained in such a climate.

Question 15 (5 marks) ©NESA 2021 SII Q32 ●●●

The flow chart shows negative feedback by the hormones testosterone and inhibin in a human male.

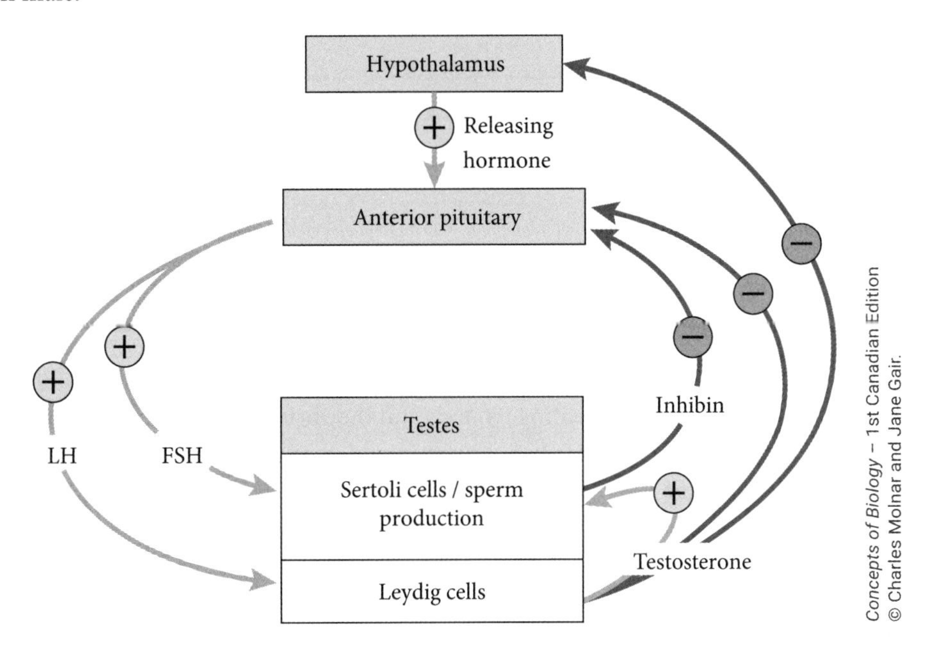

Concepts of Biology – 1st Canadian Edition © Charles Molnar and Jane Gair.

Some athletes take anabolic steroids to increase their muscle mass and strength. These steroids may be testosterone or a synthetic modification of testosterone.

Explain the changes that would occur in the testes of a male athlete continuously taking anabolic steroids. Support your answer with reference to the flow chart.

9780170465250

Test 14: Causes and effects

Section I: 10 marks. Section II: 25 marks. Total marks: 35.
Suggested time: 65 minutes

Section I: Multiple-choice questions

Instructions to students

- For each question, circle the multiple-choice letter to indicate your answer.

Question 1

Prior to the late 18th century, it was common for sailors to develop a disease called scurvy. Sailors with scurvy had swollen gums, loose teeth, slow-healing wounds and they bruised easily. In 1747, James Lind conducted an experiment on board the HMS *Salisbury* in which he subjected sailors to a variety of treatments. He found that after one week, those given oranges and lemons were well enough to nurse other patients. He also observed that those who ate rats on board the ship did not seem to develop scurvy.

What conclusion should be drawn about the cause of scurvy?

A It is caused by exposure to the sun while at sea.

B It is a genetic condition only expressed by individuals with the scurvy gene.

C It is a zoonotic disease that is passed from rats on the ships to humans.

D It is caused by a nutritional deficiency.

Question 2

Which of the following is **not** a cause of cancer?

A Some viruses

B Presence of particular genes

C Nutritional deficiencies

D Spontaneous, unregulated cell division

Question 3

The graphs below show the incidence and mortality rates of colorectal cancer for Indigenous and non-Indigenous Australians.

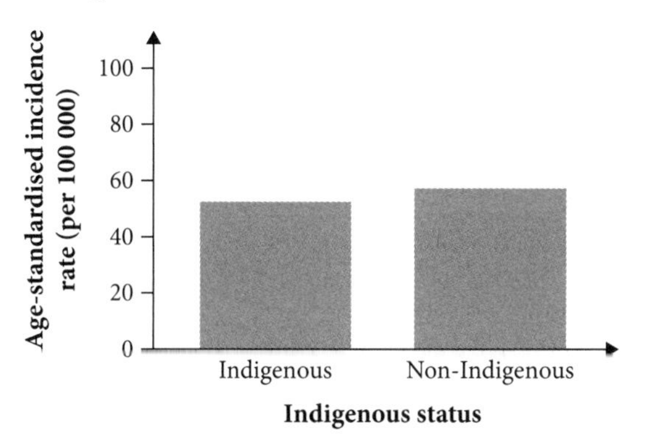

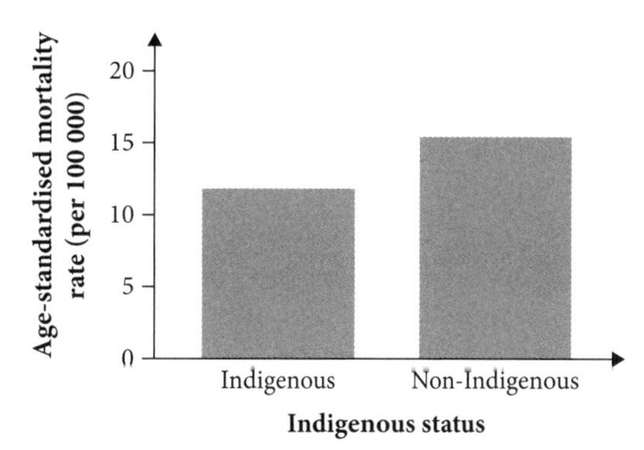

What conclusion can be drawn from these graphs?

A Indigenous Australians are less likely to die from colorectal cancer despite being more likely to develop it.

B Non-Indigenous Australians are more likely to die from colorectal cancer than Indigenous Australians.

C The chance of developing colorectal cancer is about equal in Indigenous and non-Indigenous Australians, but non-Indigenous Australians are more likely to survive.

D The chance of developing colorectal cancer is higher in non-Indigenous Australians, but the chance of surviving is approximately equal.

Question 4

The number of deaths due to a particular disease in a specified time period is called the

A incidence rate.

B prevalence rate.

C morbidity rate.

D mortality rate.

Question 5

Mesothelioma is a disease that occurs as a result of inhalation of asbestos fibres, which can cause a genetic change in the cells of the lungs, leading to cancerous growth.

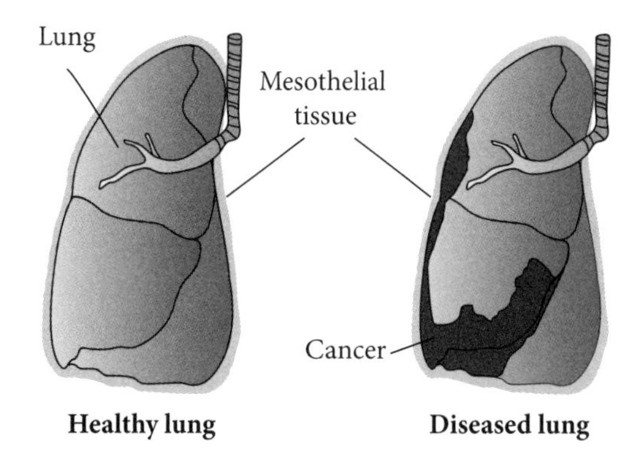

What type of disease is mesothelioma?

A Nutritional

B Genetic

C Psychological

D Environmental

Question 6

A student took the following notes during class.

- Down syndrome is caused by non-disjunction of chromosome pair 21 during meiosis, resulting in three copies of the chromosome.
- Cystic fibrosis is an inherited disease that results in an excessively thick, sticky mucus and often leads to recurrent lung infections.
- Beri beri is caused by a deficiency of vitamin B_1 in the body.
- Rickets is caused by vitamin D deficiency due to inadequate vitamin D in the body or inadequate exposure to sunlight, preventing the body from synthesising vitamin D.

When the student looked over his notes that evening, he was confused as to which category each disease should be placed in, so he checked with his peers, but found that they had all categorised them differently.

Which student A–D had correctly categorised the four diseases?

A

Nutritional	Environmental	Genetic
Rickets Beri beri	Rickets	Down syndrome Cystic fibrosis

B

Environmental	Nutritional	Genetic
Rickets Down syndrome	Rickets Beri beri	Cystic fibrosis

C

Infectious	Environmental	Genetic
Cystic fibrosis	Rickets Beri beri	Down syndrome

D

Genetic	Nutritional	Infectious
Rickets Cystic fibrosis	Beri beri	Down syndrome

Use the information below to answer Questions 7 and 8.

The graph shows the leading causes of death in 2015, along with the projected numbers for each of these diseases in 2030.

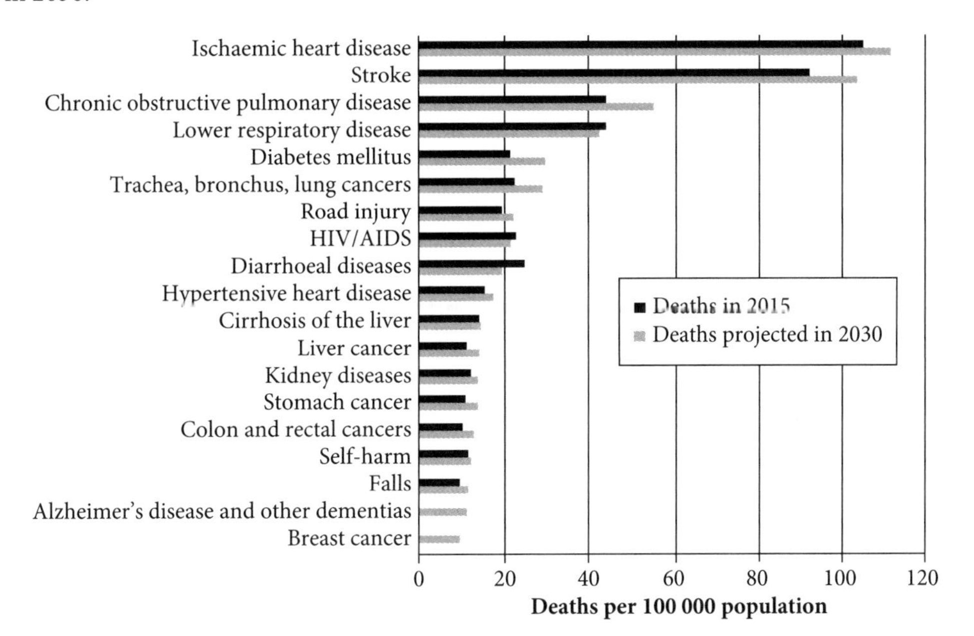

WHO, Projections of mortality and causes of death, 2015 and 2030

Question 7

What conclusion can be drawn about the predicted changes in the causes of death from 2015 to 2030?

A There will be an increase in the rate of death from infectious diseases.

B There will be an increase in the rate of death from non-infectious diseases.

C There is no clear trend that can be identified in this data.

D There will be no change to the rates of death, but the number of deaths will increase as the population continues to grow.

Question 8

Which of the following categories of disease contributed the highest number of deaths in 2015?

A Viral diseases

B Nutritional diseases

C Non-infectious diseases

D Accidental death

Use the information below to answer Questions 9 and 10.

Alcohol that is consumed is broken down by the following pathway.

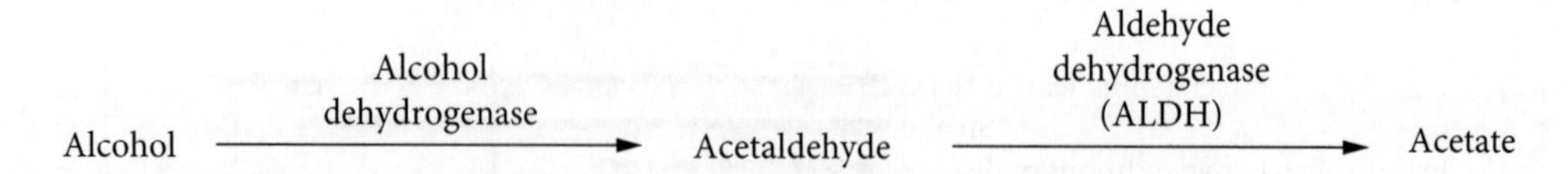

ALDH is the product of the *ALDH* gene. The normal allele is *ALDH1*, but many people have the *ALDH2* allele, which is associated with an increased risk of cancer.

Scientists have used mouse models to study the effects of *ALDH2*. They used wild-type mice with *ALDH1* alleles and genetically engineered mice with one or two copies of the *ALDH2* allele.

The mice were given a dose of alcohol and the response was measured. The results are shown.

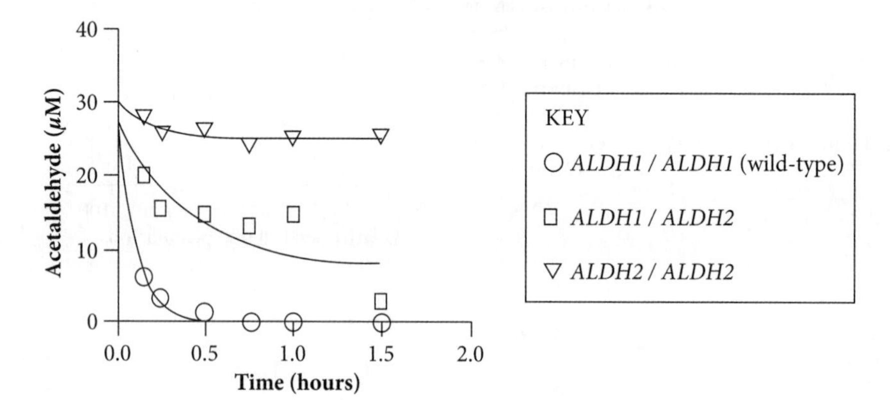

Question 9 ©NESA 2021 SI Q19

The *ALDH* alleles differ in their

A genome location.

B location in gametes.

C effect on phenotype.

D amino acid composition.

Question 10 ©NESA 2021 SI Q20

What do the data show about the effect of the *ALDH2* allele?

A Enzyme activity is highest in homozygous mice.

B Enzyme activity decreases if *ALDH2* is present.

C Enzyme activity increases if *ALDH2* is present.

D Enzyme activity is lowest in wild-type mice.

Section II: Short-answer questions

Instructions to students
- Answer all questions in the spaces provided.

Question 11 (4 marks) 

Complete the table to show the causes and effects of non-infectious diseases in humans.

Cause	Effect	Named disease with this cause
Genetic		
Environmental		

Question 12 (5 marks)

The town of Kingrey had a population of 10 174 on 1 January 2021. Data collected the previous year indicated that there were 472 people living with type 2 diabetes. During 2021, a further 32 people were diagnosed with type 2 diabetes. There were 58 deaths and 120 births in the town.

a Calculate the incidence of type 2 diabetes for Kingrey in 2021. Show all working. 2 marks

b Calculate the prevalence of type 2 diabetes on 1 January 2022. Show all working. 3 marks

Question 13 (5 marks)

Mercury occurs naturally in the environment and can accumulate in the cells of fish. Fish that are higher in the food chain, such as shark, tuna, mackerel and swordfish, are more likely to contain high levels of mercury. Pregnant women are advised to avoid eating these fish because the mercury can pass through the placenta to the baby, causing delayed development in the baby's early years and damage to the nervous system.

a Discuss whether the condition described above has a nutritional, environmental or genetic cause. 3 marks

b Is this condition infectious or non-infectious? Explain your answer. 2 marks

Question 14 (4 marks) ©NESA 2001 SII Q27

Epidemiological studies have demonstrated a relationship between ultraviolet radiation and the development of melanoma, a type of skin cancer.

The graph shows the rate of occurrence of melanoma in males and females between 1972 and 1997.

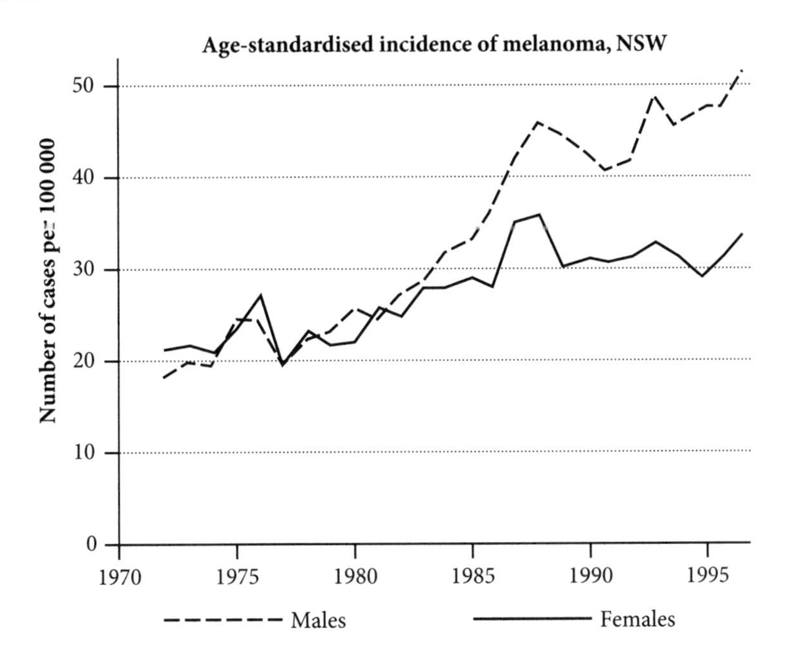

A student studying the graph made the following statement.

'The incidence of melanoma will continue to increase beyond 1997 at a greater rate in males than in females.'

Analyse the data in the graph to assess the validity of this statement.

Question 15 (7 marks)

Calcium is a mineral that makes up the structure of teeth and bones. It is also essential for keeping the body's tissues flexible and strong, and ensuring that the blood clots normally. Calcium can be obtained from dairy products, leafy green vegetables, soy products and fish bones.

It is recommended that 700 mg of calcium be consumed each day.

The data in the table indicates the average calcium intake per day for males and females in various age groups.

Age (years)	Calcium consumed (mg day^{-1})	
	Males	Females
1–9	840	810
10–24	1050	710
25–39	780	590
40–54	750	570
55–69	680	520
>70	570	450

a Plot the data in the table on the grid provided. 4 marks

b Analyse the trends shown in the graph you have presented. 3 marks

Test 15: Epidemiology

Section I: 10 marks. Section II: 25 marks. Total marks: 35.
Suggested time: 65 minutes

Section I: Multiple-choice questions

Instructions to students
- For each question, circle the multiple-choice letter to indicate your answer

Question 1

Epidemiology was defined by the World Health Organization as 'the study of the distribution and determinants of health-related states and events'.

Which of the following health-related states and events can epidemiology be applied to?

A Infectious disease only

B Non-infectious disease only

C Infectious and non-infectious disease, but not accidents

D Infectious and non-infectious disease, as well as accidents

Question 2 ©NESA 2020 SI Q8

Quarantine is ineffective as a measure to control non-infectious diseases because they

A cannot develop in isolation.

B depend on long-term exposure to a pathogen.

C may be inherited and affect the organism all their life.

D may only be treated by genetic engineering to alter cells.

Question 3

Which of the following is a random error that could affect the validity of an epidemiological study?

A A selectively chosen sample that does not represent the population to be studied

B An unpredictable variation in the data that leads to inconsistent results

C A sample that only includes current cases in the study, rather than including those who have recovered or died

D A sample that includes people who were misclassified at the start of the study and already have the condition

9780170465250

Question 4

The graph below shows the incidence of diabetes by age and gender, for total physical activity levels.

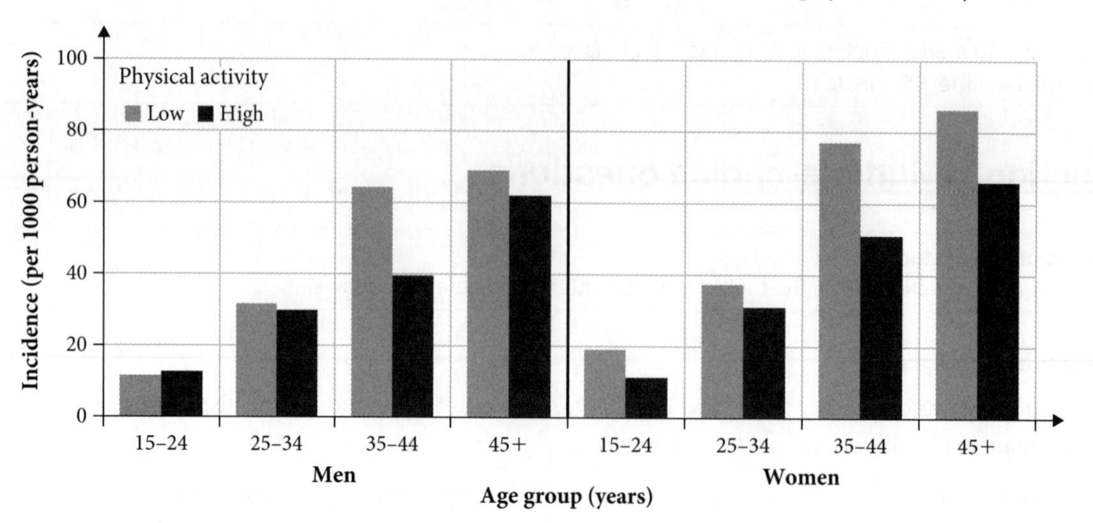

Which analysis of the data is correct?

A Physical activity has no effect on the incidence of diabetes.

B Physical activity causes diabetes in males aged 15–24.

C As people age, they engage in more physical activity than before.

D The incidence of diabetes is influenced more by age than physical activity.

Question 5

In 1947, Dr Richard Doll carried out an epidemiological study in London hospitals, comparing patients with lung cancer with those with other conditions. He collected and analysed information about their lifestyles, including smoking habits. In the same year, Dr A.B. Hill conducted a study in England, where he sent surveys to 40 000 doctors over a 10-year period. One group of doctors were smokers and the other group were non-smokers.

Which row of the table below correctly classifies the epidemiological studies of Doll and Hill?

	Doll	Hill
A	Case control	Cohort
B	Cohort	Randomised control
C	Randomised control	Case control
D	Case control	Randomised control

Question 6

The table shows the number of new cases of colorectal cancer in Australia from 1985 to 2015.

Year	Males	Females	Total
1985	780	768	1548
1995	1382	1221	2603
2005	2846	1822	4668
2015	4212	1835	6047

What trend can be seen in the number of cases of colorectal cancer in Australia?

A The rate of colorectal cancer has always been much higher in males than females.

B The rate of colorectal cancer is continuing to rise in males but has begun to stabilise in females.

C The total number of cases of colorectal cancer is starting to decline.

D There are more males in the total population, so a higher number of cases does not indicate a higher rate of colorectal cancer.

Question 7

A pharmaceutical company wishes to test the effectiveness of a new medication to treat Alzheimer's disease.

What type of epidemiological study would be used?

A An intervention study

B A cohort study

C A descriptive study

D An analytical study

Question 8

The graph shows the incidence and prevalence of melanoma in 2012.

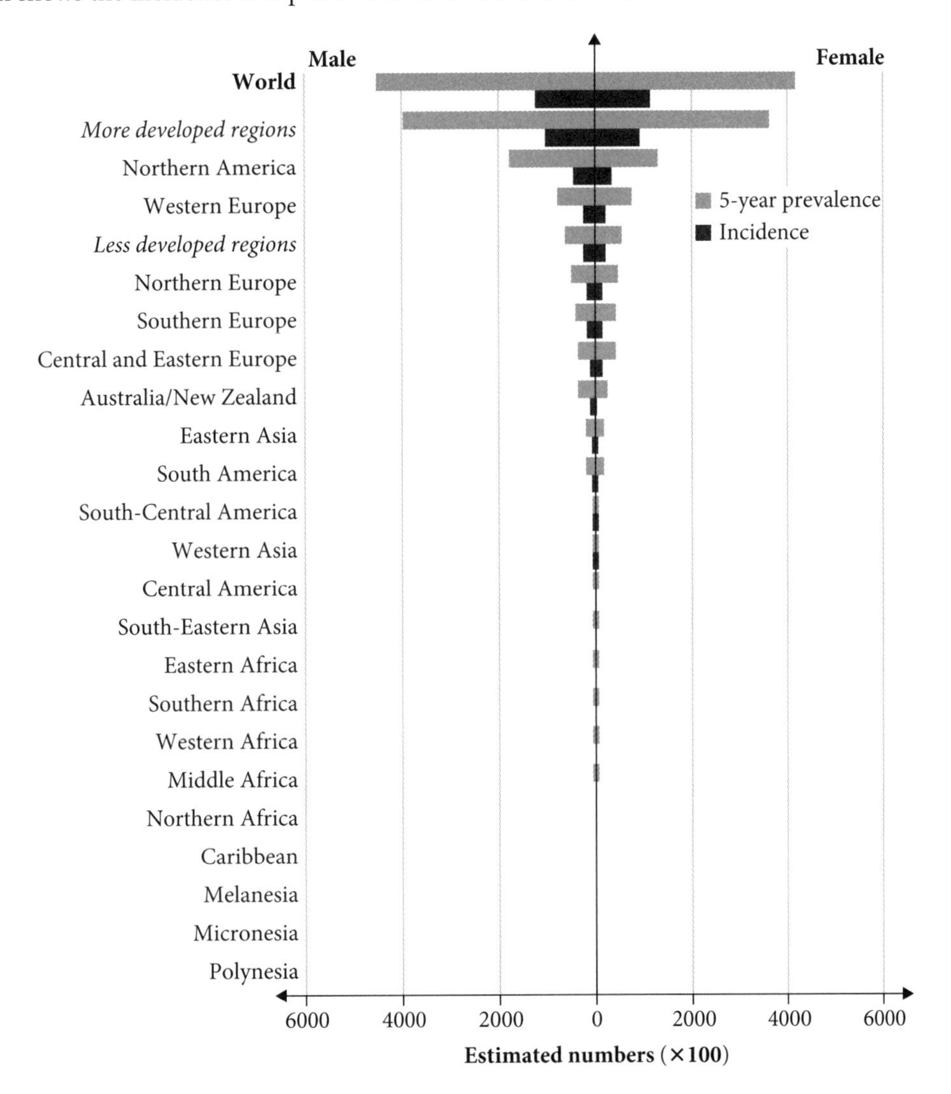

Ferlay J., Soerjomataram I., Ervik M., Dikshit R., Eser S., Mathers C., Rebelo M., Parkin D.M., Forman D., Bray F. GLOBOCAN 2012 v1.0, Cancer Incidence and Mortality Worldwide: IARC CancerBase No. 11 [Internet]. Lyon, France: International Agency for Research on Cancer; 2013. Available from: http://globocan.iarc.fr, accessed on February 2017.

Which statement is correct?

A The worldwide incidence of melanoma in 2012 was higher in females than in males.

B The incidence of melanoma is higher in less-developed regions, but the prevalence is higher in more-developed regions.

C The incidence and prevalence of melanoma is higher in more-developed regions of the world.

D People in Northern Africa and Polynesia cannot develop melanoma.

9780170465250

Question 9 ©NESA 2021 SI Q4

Which is the most effective strategy for treating non-infectious diseases?

A Hygiene

B Pharmaceuticals

C Quarantine

D Vaccination

Question 10

The graph below shows the incidence rate for all cancers combined in Australia from 1982 to 2016.

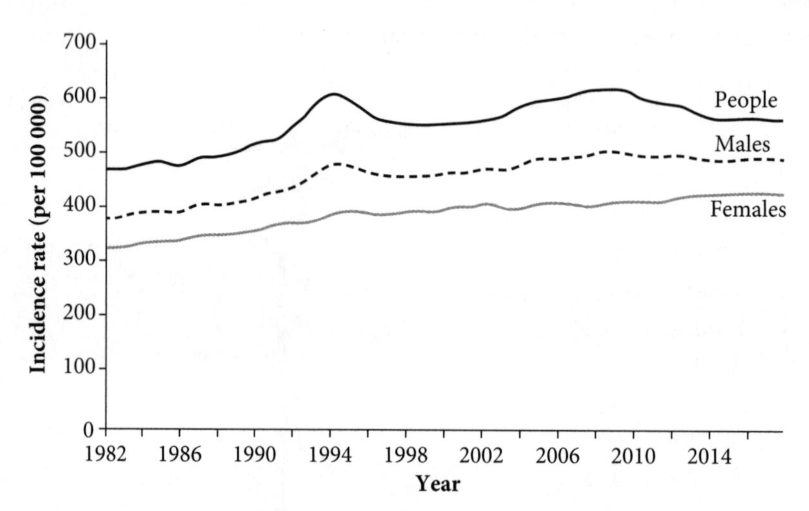

Data sourced from AIHW Cancer Data in Australia 2022 web report and supplementary data tables

What conclusion could be drawn from this data?

A The rate of cancer is increasing in males but decreasing in females.

B The rate of cancer is decreasing, but the number of cases is increasing as the population grows.

C The rate of all cancers increased from 1982 to 2016.

D The rate of cancer will start to increase in females and decrease in males.

Section II: Short-answer questions

Instructions to students

- Answer all questions in the spaces provided.

Question 11 (5 marks)

A student wanted to know if regular exercise would decrease the chance of developing cardiovascular disease (CVD). She went to her local gym and surveyed 24 interested members (19 females and 5 males). She asked each member how many times a week they exercised and whether they had CVD.

Her results indicated that of the 24 people surveyed, only two had CVD. She concluded that regular exercise decreased the risk of developing CVD.

Evaluate the study conducted by this student and propose changes that could be made.

Question 12 (4 marks)

Discuss the treatment, management and directions of future research for a named non-infectious disease.

Question 13 (6 marks) ©NESA 2021 SII Q31 ●●

Millions of people around the world take drugs known as statins, which have been shown to reduce the incidence of heart attacks and strokes in vulnerable patients. However, up to 20% of people stop taking statins because of side-effects such as muscle aches, fatigue, feeling sick and joint pain.

A recent study at a public hospital focused on 60 patients who had all stopped taking statins in the past due to severe side-effects. Patients took statin tablets for four months, placebo tablets for four months and no tablets for four months.

Every day for the year the patients scored, from zero to 100, how bad their symptoms were. The results are shown.

Four month treatment	Average score symptoms / 100
Statin tablets	16.3
Placebo tablets	15.4
No tablets	8.0

Evaluate this study and its results.

Question 14 (7 marks)

In 2014, a World Health Organization report determined that there was evidence that consuming red meat and processed meat products could significantly increase the risk of developing colorectal cancer. The risk appears to increase with the amount of red meat and processed meat products consumed.

a Design an epidemiological study that could have been used to determine the link between eating red meat and processed meat products and developing colorectal cancer. 5 marks

b Outline the benefits of conducting an epidemiological study. 2 marks

Question 15 (7 marks)

Protein is an essential macronutrient that builds cell structure, allowing for growth and repair of the body, and produces hormones, enzymes and antibodies for immunity. Protein can be obtained through a healthy diet from sources including lean meat, fish, eggs, beans and legumes. Many individuals do not consume the recommended amount of protein per day.

In some low-income countries where people live in large cities, there is a growing incidence of obesity, which may be attributed in part to excessive carbohydrate intake and low protein intake.

The graph below shows the average amount of protein consumed per capita for the world's population, high-income countries and low-income countries.

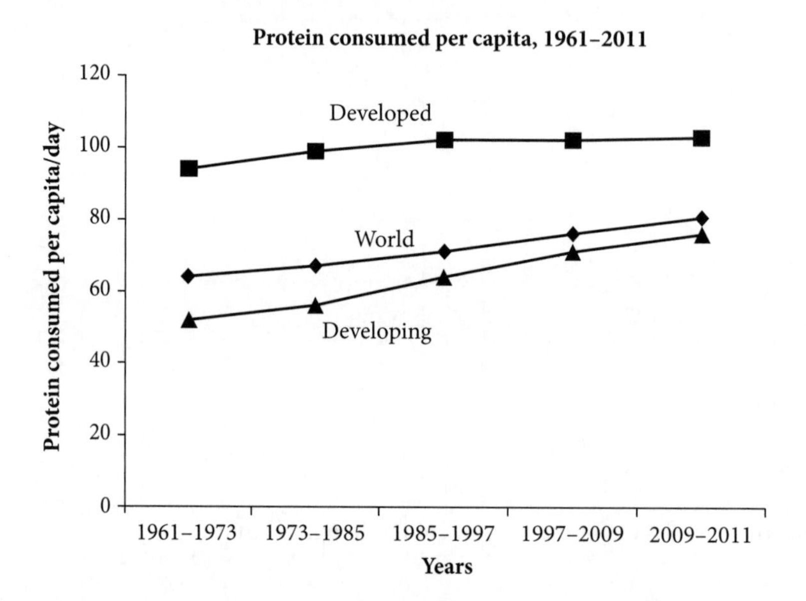

The graph below shows the price of food sources per energy unit.

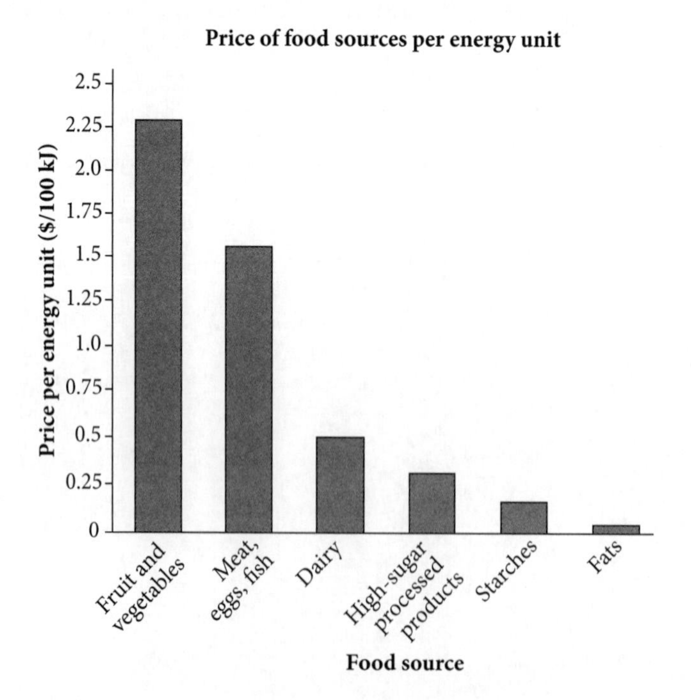

Analyse the data presented in the graphs and propose a reason for the observed trends in the incidence of obesity in some developing countries.

Test 16: Prevention

Section I: 5 marks. Section II: 13 marks. Total marks: 18.
Suggested time: 40 minutes

Section I: Multiple-choice questions

Instructions to students
- For each question, circle the multiple-choice letter to indicate your answer.

Question 1 ©NESA 2020 SI Q9 (ADAPTED)

A public education campaign was developed with the aim of lowering the incidence of skin cancer in the population.

The campaign was adopted Australia wide and is illustrated in the poster.

Cancer Council Australia

How effective would this campaign be in reducing the rate of skin cancer in Australia?

A It would be quite effective because it suggests simple ways to avoid a known risk factor, encouraging the population to change their behaviour.

B It would be highly effective because it provides legislation that must be followed to prevent skin cancer.

C It would not be effective because only infectious diseases can be prevented.

D It would not be effective because it is too simple and should have a lot more information in the graphic.

Question 2

Which of the following is **not** a current use of genetic engineering for the prevention of non-infectious disease?

A Using array comparative genomic hybridisation to test for the presence of genetic conditions in embryos

B Genetic testing of babies at birth for genetic conditions such as phenylketonuria or cystic fibrosis

C Adding genes to food crops to add nutrients and help prevent nutritional deficiencies

D Genetic modification of a virus to produce vaccines that trigger an immune response without causing disease

Question 3

Vitamin A deficiency can cause blindness and compromised immunity. Vitamin A can be obtained by consuming animal products, including meat and eggs, and vegetables such as carrots, beetroot, tomatoes and leafy greens. Vitamin A deficiency is common in countries where people consume a diet consisting largely of rice or other carbohydrates.

Scientists have genetically engineered rice, known as Golden Rice, to include a gene from maize and another from a soil bacterium that causes the plant to produce vitamin A. The graph below shows the daily intake of vitamin A from regular rice in countries where rice is a staple food, compared with the projected intake if Golden Rice was introduced.

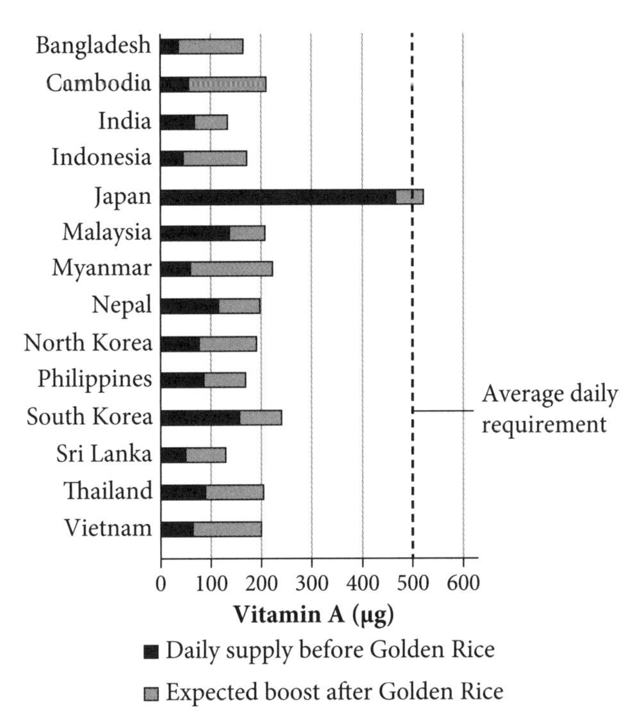

https://grain.org/article/entries/10-grains-of-delusion-golden-rice-seen-from-the-ground

What conclusion could be drawn about the introduction of Golden Rice to countries where rice is a staple food?

A Of all countries shown, Japan will have the greatest projected increase in vitamin A compared to the current intake.

B Golden Rice will allow people in all countries to consume the required daily intake of vitamin A.

C Golden Rice will make no difference to the health of people in the countries included in the graph, because it does not produce enough vitamin A to meet the daily requirement.

D The amount of vitamin A consumed will increase in all countries shown in the graph but will not allow people to reach the recommended daily intake in most of the countries.

Question 4

The National Bowel Cancer Screening Program was established in Australia in 2006. Individuals between the ages of 50 and 74 years are mailed a kit that is used to take a sample of their faeces in the privacy of their own home. The sample is then sent to a laboratory to test for the presence of blood in the faeces, which may be an indicator of bowel cancer.

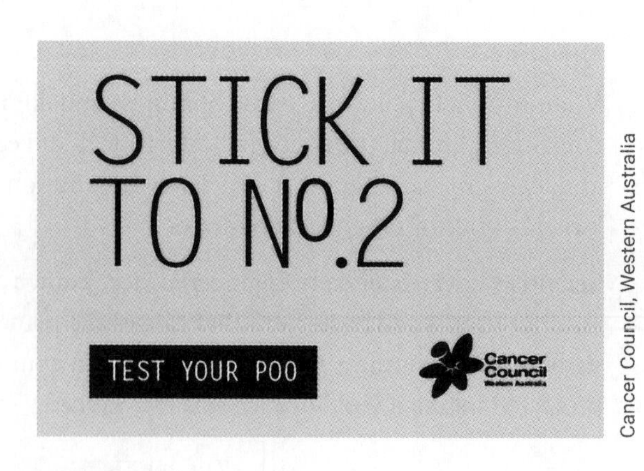

Cancer Council, Western Australia

How could the effectiveness of the National Bowel Cancer Screening Program be measured?

A By the reduction in incidence and prevalence of all types of cancers

B By the increase in sales of insoluble fibre products that claim to reduce the risk of bowel cancer

C By the increase in the number of people aged 50–74 participating in the screening program

D By surveying people aged 50–74 and asking them whether they thought the campaign was effective

Question 5

What role can governments play in preventing non-infectious disease?

A A government's role is limited to passing legislation. Genetic engineering and education campaigns are managed by private organisations.

B Governments are responsible for passing legislation and developing education programs. Genetic engineering is the concern of private organisations only.

C Governments are responsible for passing legislation as well as funding research, regulating genetic engineering and developing public education campaigns.

D Governments do not play a role in preventing non-infectious diseases because these are caused by lifestyle choices, which are the concern of the private sector.

Section II: Short-answer questions

Instructions to students

- Answer all questions in the spaces provided.

Question 6 (4 marks)

Using examples, outline the features of an effective educational program or campaign in preventing non-infectious disease.

Question 7 (4 marks)

In addition to increasing the price of cigarettes, the Australian Government has implemented legislation aimed at decreasing the burden of smoking on health in Australia. The graph below shows the prevalence of smoking in Australia from 1990 to 2015, highlighting the dates when key pieces of legislation were introduced.

Assess the role of legislation in reducing lung cancer in Australia.

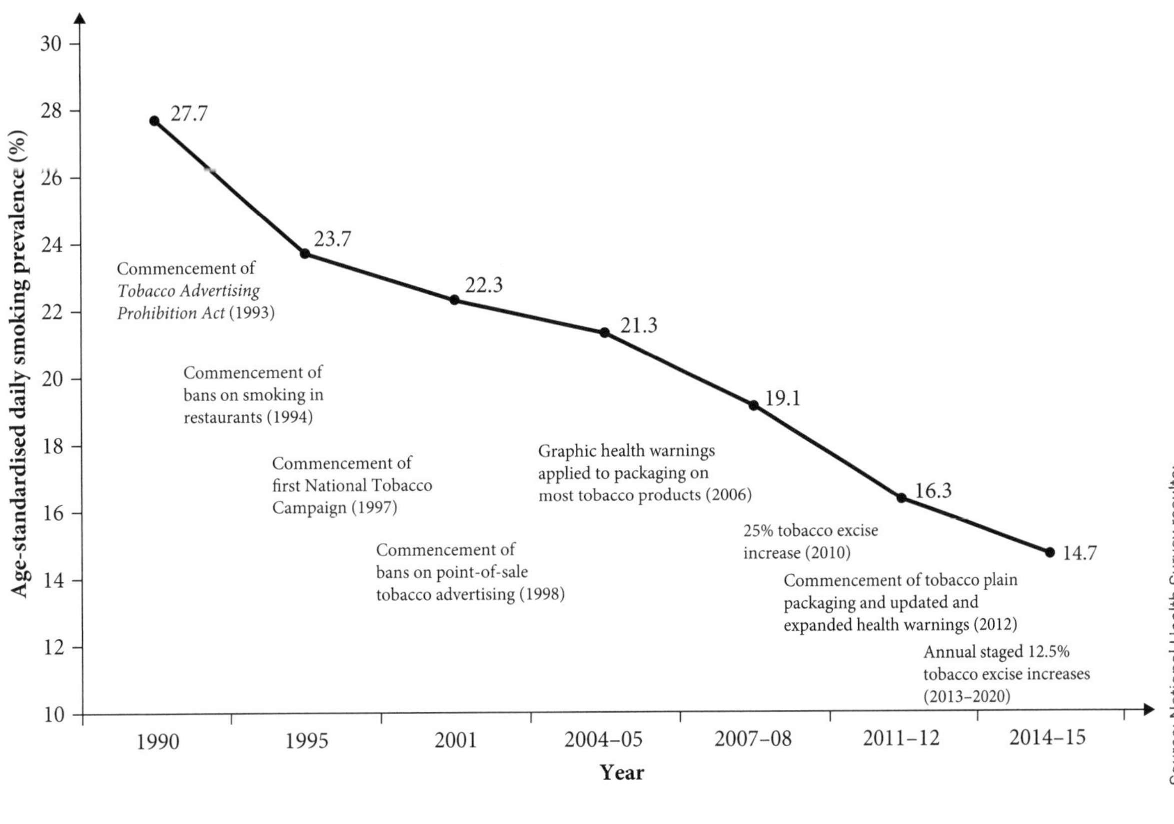

Source: National Health Survey results: Australian Bureau of Statistics (4364.0.55.001) Creative Commons Attribution 2.5 Australia licence.

Question 8 (5 marks)

Sickle-cell anaemia affects the production of haemoglobin, resulting in cells that are crescent-shaped rather than disc-shaped.

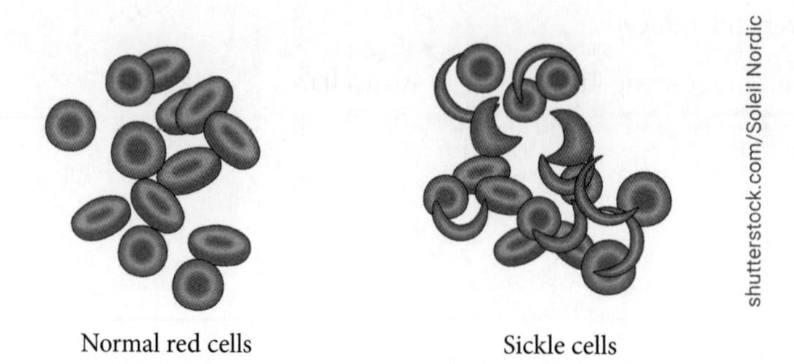

Crescent-shaped cells cannot carry oxygen as efficiently as normal cells and can block blood flow, leading to stroke, intense pain and death. Medication and blood transfusions have been used to lessen the severity of sickle-cell anaemia, but there is still no cure.

In adults, a haemoglobin molecule is made up of two α-globin units and two β-globin units. In a person with sickle-cell anaemia, a mutation in the β-globin prevents effective transport of oxygen in the blood. In a foetus, haemoglobin consists of two α-globin units and two γ-globin units. The γ-globin can transport oxygen but is usually inactivated at birth. A mutated gene exists in some people that allows the γ-globin to remain active.

In 2019, scientists collected stem cells from two women with sickle-cell anaemia and used CRISPR-Cas9 to edit the gene that normally inhibits γ-globin production in adults, so that the gene no longer acts as an inhibitor. The stem cells were then reintroduced to the patients' bone marrow, where they would differentiate to form red blood cells containing foetal haemoglobin (α- and γ-globin).

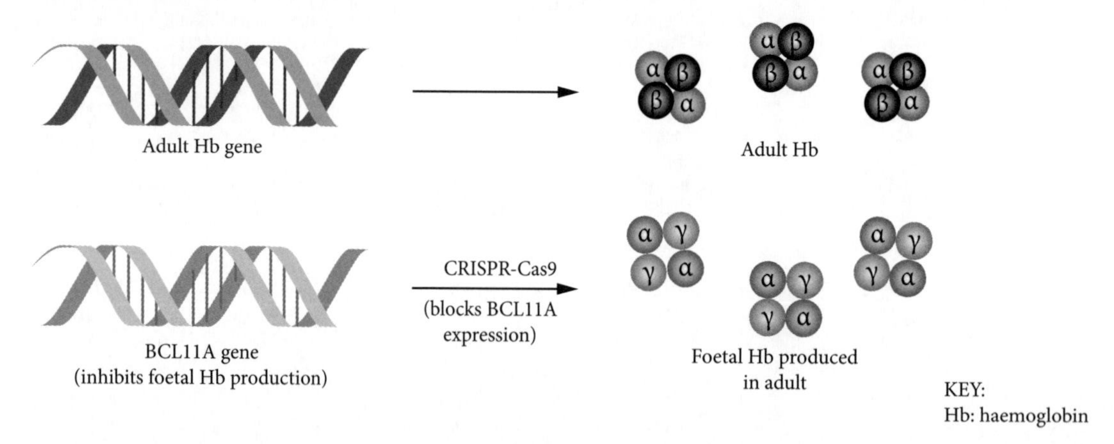

Five months after this procedure, one patient had foetal haemoglobin in 98% of her red blood cells, and the other had foetal haemoglobin in 94% of red blood cells. Neither had experienced any pain episodes since receiving treatment, and no side-effects were reported.

Evaluate the effectiveness of the genetic engineering technology CRISPR-Cas9 in preventing sickle-cell anaemia.

Test 17: Technologies

Section I: 10 marks. Section II: 30 marks. Total marks: 40.
Suggested time: 70 minutes

Section I: Multiple-choice questions

Instructions to students

- For each question, circle the multiple-choice letter to indicate your answer.

Question 1 ©NESA 2019 SI Q6

How does the cochlear implant assist people with severe hearing loss?

A It amplifies sound.

B It stimulates the ear drum.

C It stimulates the auditory nerve.

D It amplifies vibrations in the cochlea.

Question 2 ©NESA 2016 SI Q7

The diagram shows a simplified model of a mammalian nephron and the three processes labelled *1*, *2* and *3*.

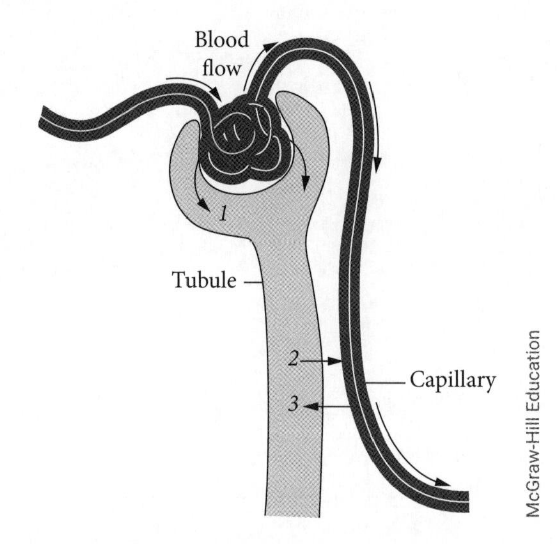

Which row of the table correctly identifies each of the processes?

	1	*2*	*3*
A	Reabsorption	Filtration	Secretion
B	Filtration	Reabsorption	Secretion
C	Secretion	Reabsorption	Filtration
D	Filtration	Secretion	Reabsorption

Question 3

Dialysis treatment saves the lives of thousands of people with kidney disorders each year. However, it can only carry out some of the processes of the nephron.

Which of the following processes can dialysis achieve?

A Filtration only

B Reabsorption only

C Both filtration and reabsorption

D Secretion only

Use the following diagram to answer Questions 4 and 5. The diagram shows the structures of the human ear.

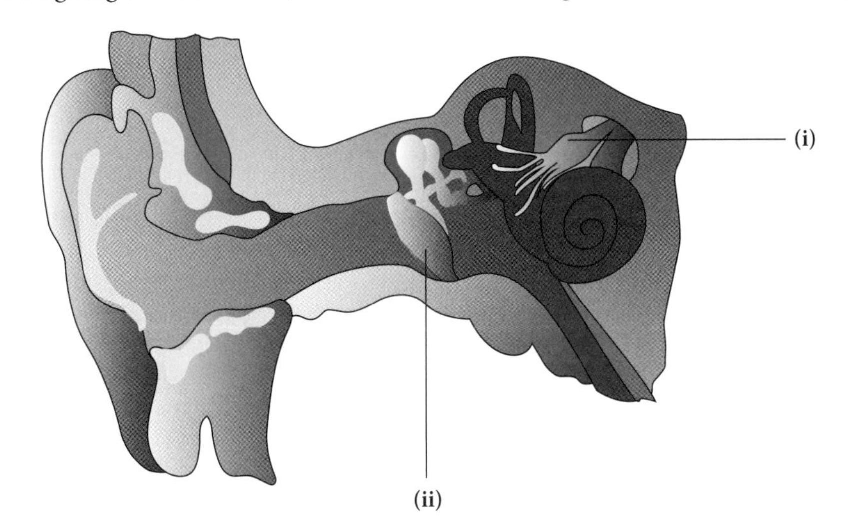

Question 4

What is the name of the structure labelled (i) in the diagram?

A Cochlea

B Vestibular canals

C Ossicles

D Auditory nerve

Question 5

What is the function of the structure labelled (ii) in the diagram?

A To transmit and amplify sound to the inner ear

B To aid with balance of the body

C To convert mechanical waves to electrical energy

D To send electrical impulses to the brain

Question 6

The definition of 'accommodation' in the mammalian eye is the

A contraction and relaxation of the iris muscles to change the size of the pupil.

B reshaping of the cornea surface using laser technology to change its refractive power.

C relaxation of the ciliary muscles to change the shape of the lens.

D ability of the lens to change shape in order to focus light at varying distances.

Question 7

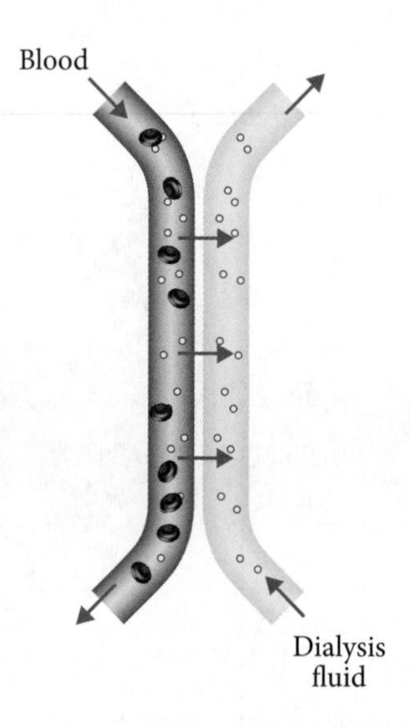

What is the function of the counter-current flow of the blood and the filtrate in dialysis?

A It maintains the concentration gradient of the blood and the dialysis fluid, to speed up diffusion of waste.

B It maintains the same concentration of water, glucose and salt on either side of the membrane.

C It ensures that there are no air bubbles in the blood that is returned to the body.

D It ensures that the temperature of the blood and the dialysis fluid is kept constant.

Question 8

What visual disorder could be treated using intraocular lens implantation?

A Myopia

B Hyperopia

C Cataracts

D Macular degeneration

Question 9

The graph shows the age at which hearing loss begins in males and females.

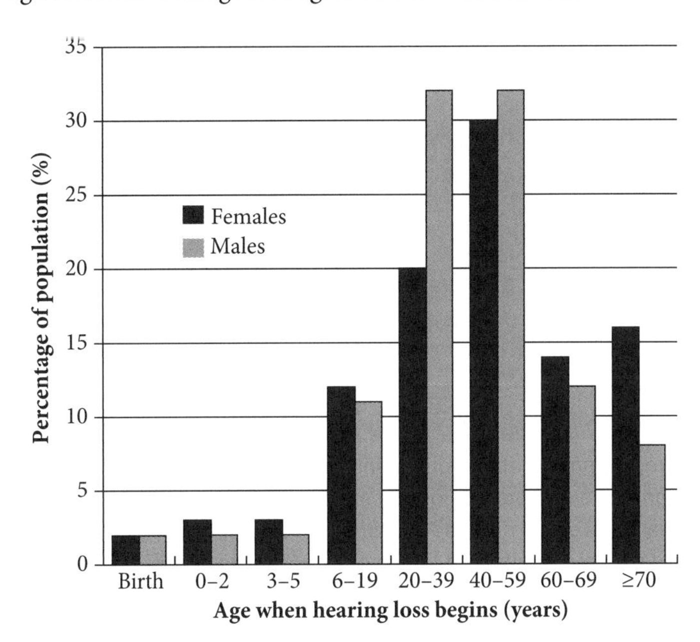

Source: National Health Interview Survey, 2007 (CDC).

What conclusion could be drawn from the data in the graph?

A Males are at an increased risk of hearing loss due to genetic causes.

B Males are not likely to lose their hearing after the age of 70.

C Males between the ages of 20 and 60 are more likely to lose their hearing than females.

D Males are at an increased risk of hearing loss due to environmental causes.

Question 10

Technologies to assist with hearing loss can be limited in their usefulness, depending on the severity of the hearing loss. The graph below shows familiar sounds that can be heard by a person with profound hearing loss unaided, with a hearing aid and with a cochlear implant. The lines indicate the minimum volume for each frequency at which the person would be able to hear the sound.

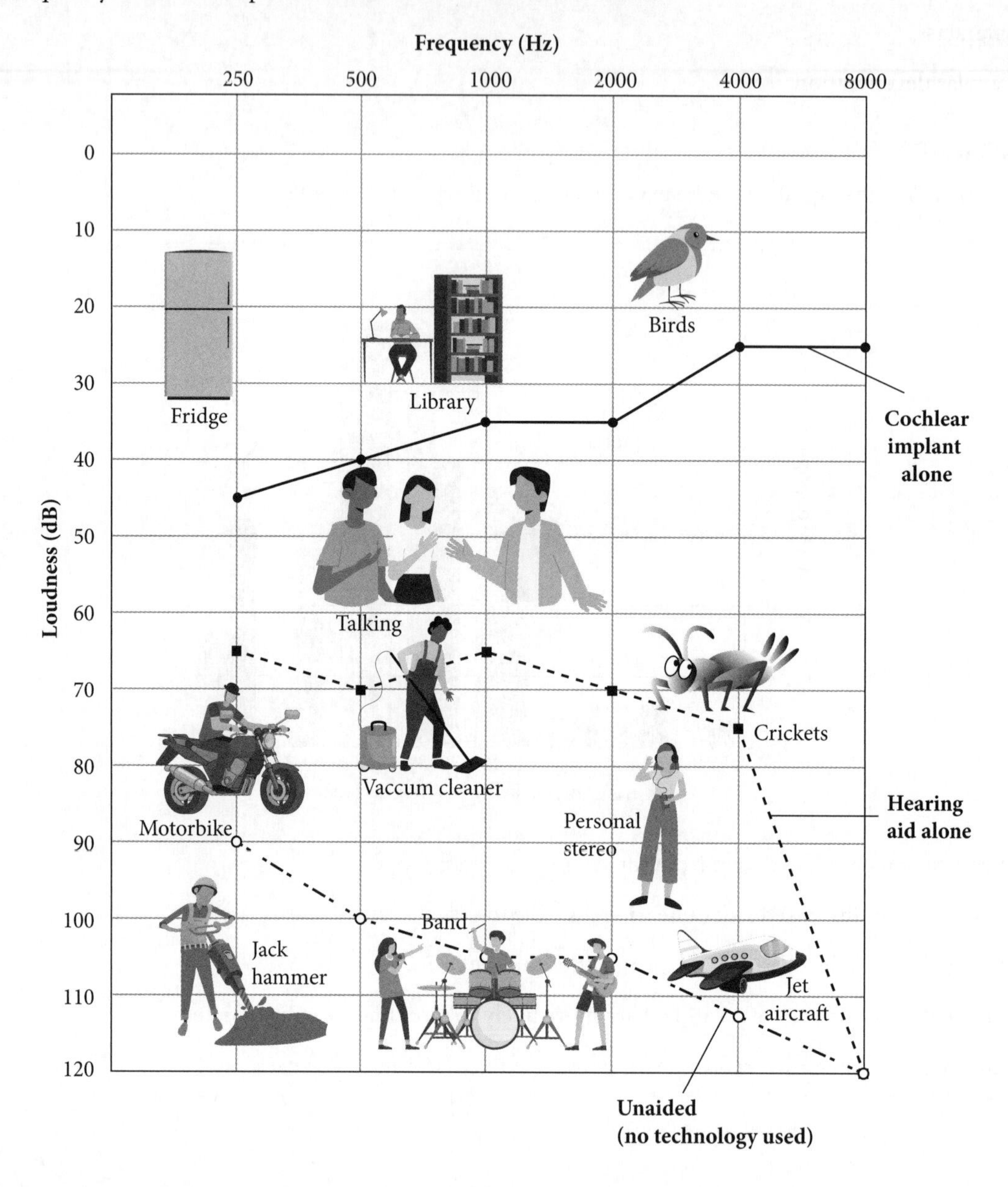

Which statement is correct?

A A person with a cochlear implant can hear sounds quieter than 20 dB if the frequency is high.

B A person with profound hearing loss can only hear loud sounds of high frequency.

C A person with a cochlear implant can hear most frequencies required for conversation, as long as the sound is loud enough.

D A person with profound hearing loss can hear higher frequencies with a hearing aid than with a cochlear implant.

Section II: Short-answer questions

Instructions to students
- Answer all questions in the spaces provided.

Question 11 (3 marks)

Complete the table below, identifying the name and function of each structure labelled in the diagram.

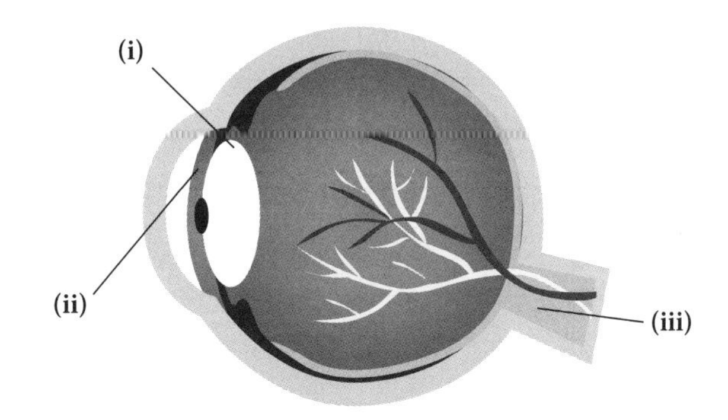

Structure	Name	Function
(i)		
(ii)		
(iii)		

Question 12 (8 marks) ©NESA 2021 SII Q25

A patient visited an audiologist for a hearing test. The audiologist tested both ears at specific frequencies. The volumes at which each frequency could be heard are shown.

Frequency (Hz)	Minimum volume at which sound could be detected (dB)	
	Right ear	Left ear
250	5	55
500	9	60
1000	9	75
2000	5	75
4000	9	80
8000	20	100

a Plot the data on the grid provided and include a key. 3 marks

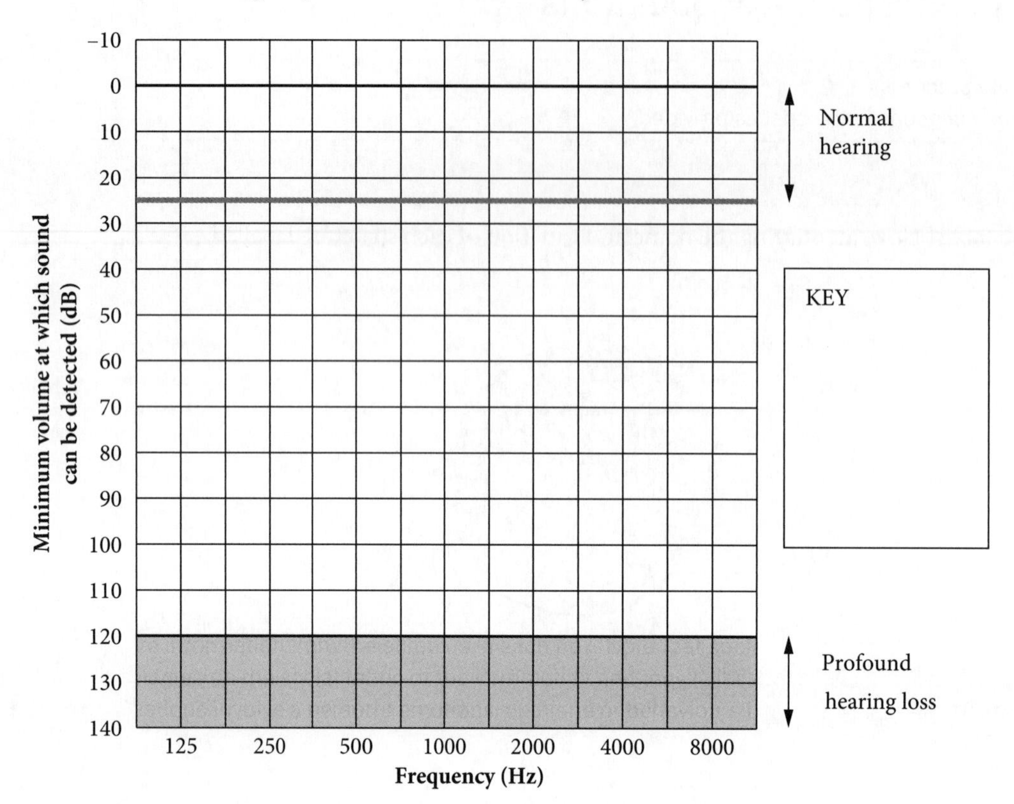

b What conclusions can be drawn about the patient's hearing? 2 marks

c It is discovered that there is a complete permanent blockage of the outer ear, but the cochlea is still fully functional.

Justify the use of a suitable technology to assist the patient's hearing. 3 marks

Question 13 (9 marks)

a Describe the role of the kidney in maintaining human health. 2 marks

b Outline the cause and effects of a named kidney disorder. 3 marks

c A person suffering from kidney failure will need to undergo dialysis treatment and, in some cases, may receive a kidney transplant. Evaluate the effectiveness of each of these technologies in managing the effects of a kidney disorder. 4 marks

Question 14 (6 marks)

a With the use of a diagram, explain the cause of a named visual disorder that prevents formation of a focused image. 3 marks

b Describe how **one** technology can assist people who suffer from the disorder you named in part **a**. 3 marks

Question 15 (4 marks)

Hearing loss can have a variety of causes. Compare the use of hearing aids and cochlear implants to manage hearing loss. In your answer, refer to the location of the technology, conditions under which it may be beneficial, and the benefits and limitations of each technology.

Biology

PRACTICE HSC EXAM 1

General instructions

- Reading time: 5 minutes
- Working time: 3 hours
- Draw diagrams using pencil
- Calculators approved by NESA may be used

Total marks: 100

Section I – 20 marks

- Attempt Questions 1–20
- Allow about 35 minutes for this section

Section II – 80 marks

- Attempt Questions 21–31
- Allow about 2 hours and 25 minutes for this section

9780170465250

Section I

20 marks
Attempt Questions 1–20
Allow about 35 minutes for this section

Circle the correct multiple-choice option for Questions 1–20.

Question 1

Which statement correctly describes fungi and protozoa?

A Fungi and protozoa are unicellular.

B Fungi and protozoa have chloroplasts.

C Fungi have a cell wall and protozoa do not.

D Fungi are prokaryotic and protozoa are eukaryotic.

Question 2

An epidemiologist designed a study to examine the relationship between daily hours of sleep and weight gain. She initially recruited 10 people to participate in the study, but upon discussing her work with a colleague, she increased this number to 100 participants.

By increasing her sample size from 10 to 100 participants, which of the following would be improved?

A Accuracy

B Precision

C Reliability

D Validity

Question 3

A gardener noticed that a leaf which had fallen off a succulent plant had sprouted small roots and shoots, as shown in the diagram.

After several weeks, the original leaf began to shrivel. The gardener removed the leaf and planted the remaining plantlet into moist soil, where it continued to grow.

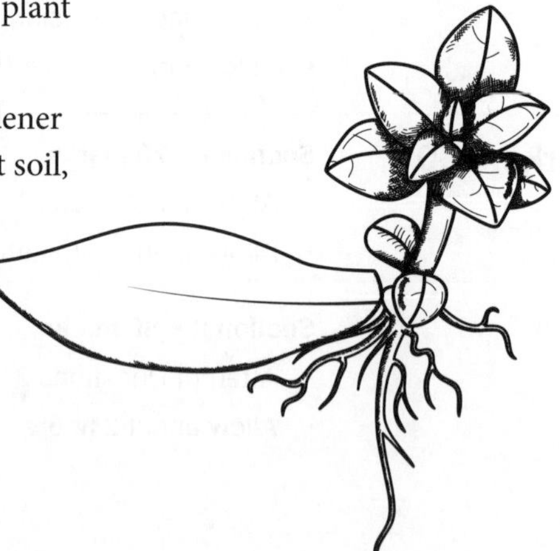

What process occurred in this example?

A Tissue culture

B Artificial pollination

C Sexual reproduction

D Vegetative propagation

Question 4

Which of the following is produced as a result of transcription?

A A polypeptide

B A DNA strand

C An amino acid

D An mRNA strand

Question 5

Australian plants have developed many adaptations that enable them to conserve water in an environment where water may be scarce. One example is marram grass, shown in the diagram below in cross-section.

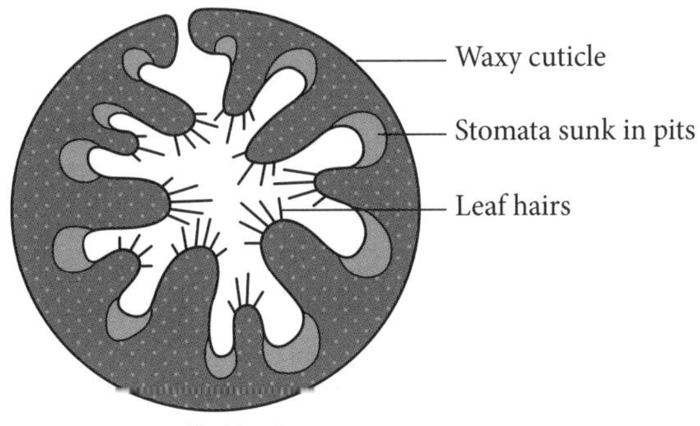

Which of the following correctly describes how one of these adaptations assists water conservation?

A The waxy cuticle reflects sunlight and reduces transpiration.

B The leaf rolls up to increase humidity and reduce transpiration.

C Sunken stomata decrease the surface area to increase transpiration.

D Leaf hairs increase surface area, leading to increased water absorption and reduced transpiration.

Question 6

The graph below shows some changes that occur during the menstrual cycle.

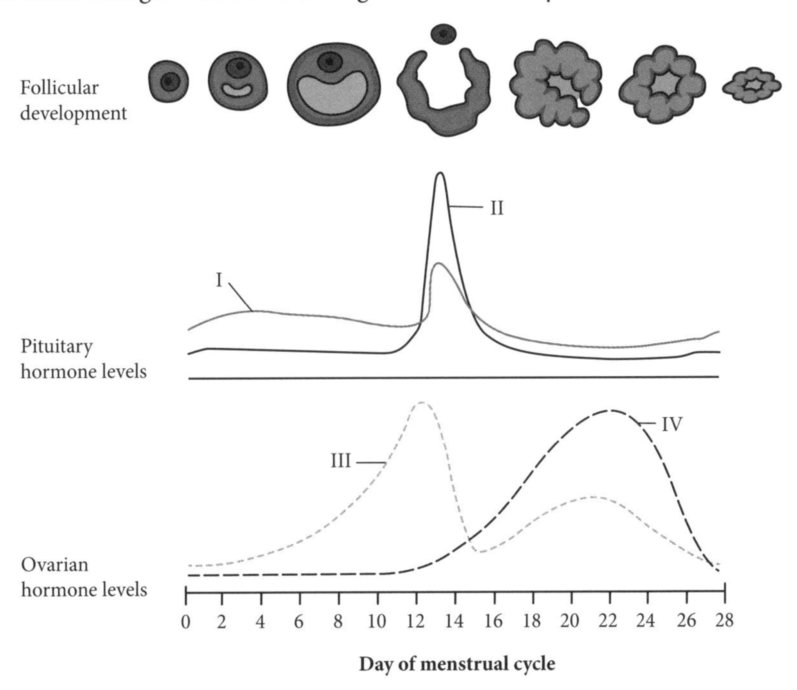

Which of the following represents the level of follicle-stimulating hormone during the menstrual cycle?

A I

B II

C III

D IV

Question 7

The diagram below shows a structure involved in polypeptide synthesis.

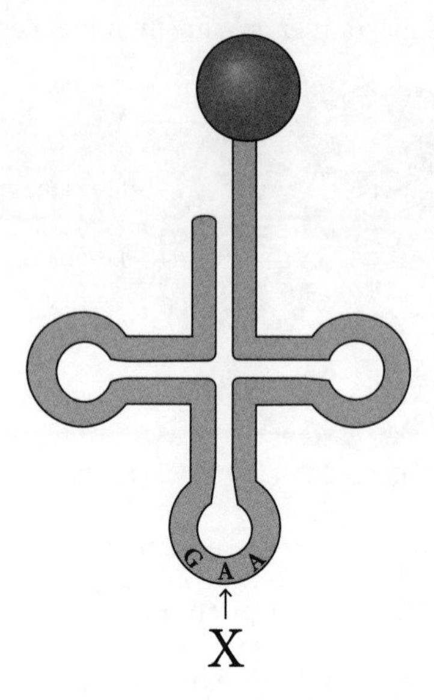

The molecule that attaches to this structure at point X would contain

A a codon with the sequence CTT.

B a codon with the sequence CUU.

C an anticodon with the sequence CTT.

D an anticodon with the sequence CUU.

Question 8

Steps in a reproductive process used to produce a horse with certain characteristics are listed below.

Step 1 – An unfertilised egg was retrieved from horse W and the nucleus was removed.

Step 2 – A body cell was taken from horse X. The nucleus was removed and inserted into the cell from Step 1.

Step 3 – The resulting cell was implanted into the uterus of horse Y.

Step 4 – Horse Y gave birth to horse Z.

Which horse(s) would be genetically most similar to horse Z?

A W

B X

C W and X

D X and Y

Question 9

Indigenous Australians have lived on this continent for tens of thousands of years and have developed a rich knowledge of the potential uses of plants for medicinal purposes. One such example is emu bush, which was used by Indigenous people in the Northern Territory to disinfect wounds and occasionally as a gargle for sore throats. Scientists have recently discovered that its leaves have the same potency as many established antibiotics, and they plan to use them to sterilise implants prior to surgery.

Emu bush

Which of the following statements about the use of bush medicines such as emu bush is correct?

A Antibiotics sourced from bush medicine are unlikely to be as effective as those from other well-known sources.

B Indigenous Australians no longer use bush medicines, so it is not necessary to consider their cultural views on intellectual property.

C Australian laws require consultation with local Aboriginal Elders before an organisation can apply for the rights to use a resource.

D Australian intellectual property laws do not align well with Indigenous Australians' cultural view that no one can exclusively own the rights to a resource.

Question 10 ©NESA 2021 SI Q10

Cystic fibrosis is an autosomal recessive disorder caused by mutations in the *CFTR* gene. Many different recessive alleles cause cystic fibrosis.

The four most common alleles of the *CFTR* gene and their frequencies in the Australian population are shown in the table.

Allele	Frequency of allele (%)
A	98.33
a1	1.13
a2	0.08
a3	0.07

What will be the most common genotype of cystic fibrosis patients in Australia?

A a1/a1

B a1/a2

C A/a1

D A/A

Question 11

Malaria is a disease caused by protozoa in the *Plasmodium* genus. It is transmitted by *Anopheles* mosquitoes and is prevalent in tropical and subtropical areas, including parts of South Africa. Scientists have previously used a pesticide called dichlorodiphenyltrichloroethane (DDT) to reduce the number of cases of malaria.

DDT use in South Africa was discontinued in 1996 due to concerns about its effects on the environment and on human health, but it was reintroduced in 2001.

The graph below shows the number of new cases and deaths from malaria in South Africa from 1995 to 2011.

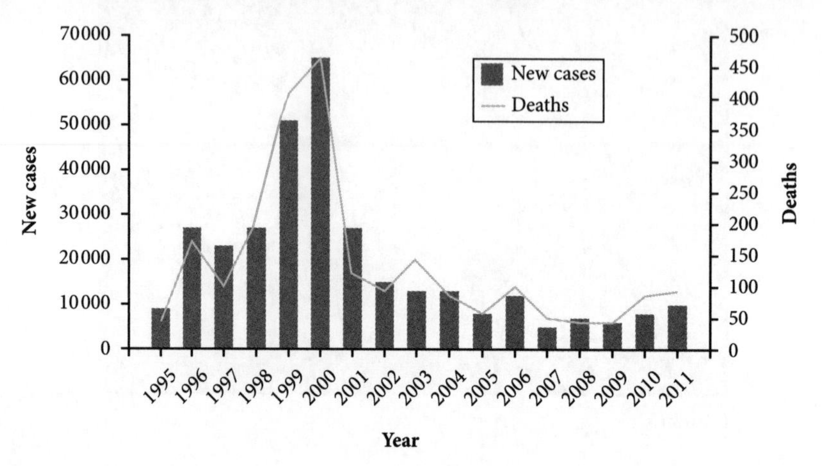

What conclusion can be drawn from this graph?

A The incidence rate of malaria was reduced due to the use of DDT.

B The prevalence rate of malaria peaked at approximately 64 000 in 2000.

C The mortality rate from malaria was higher in 2011 than the incidence rate.

D DDT is ineffective at reducing malaria rates, as there were still cases when it was used.

Question 12

DDT has been banned in many parts of the world. Short-term effects of exposure to high doses include vomiting, tremors and seizures. In the long term, DDT is thought to affect liver function and reproduction, and act as a cancer-causing mutagen.

People can be exposed to DDT through consuming food including meat, fish and dairy, through breathing or touching objects contaminated with DDT, or through breast milk passed from mother to child.

What category of disease is described above?

A Environmental

B Genetic

C Infectious

D Nutritional

Question 13

Louis Pasteur's famous swan-necked flask experiment provided evidence that disproved the theory of spontaneous generation. A scientist recreated this experiment, by boiling beef broth in three swan-necked flasks and leaving them for 30 days. After 30 days he broke the neck off one and tilted another sideways, then left all three flasks for a further 30 days, as shown in the diagram below. The flasks were then observed for signs of microbial growth.

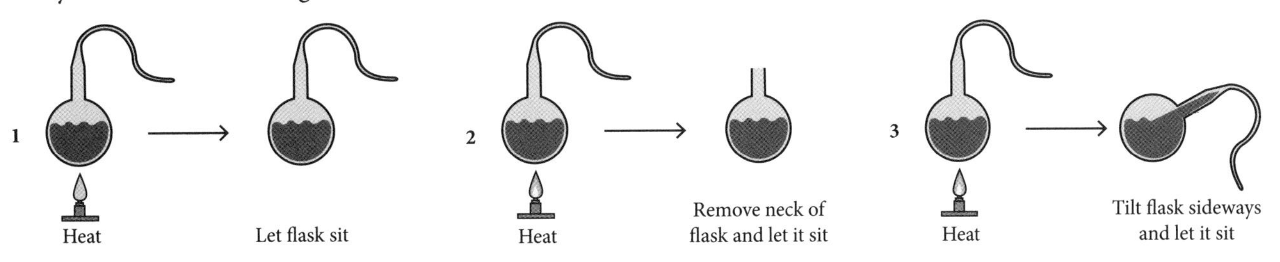

What would you expect to observe in each of the three flasks?

	Flask 1	Flask 2	Flask 3
A	Microbial growth	Microbial growth	Microbial growth
B	No microbial growth	Microbial growth	No microbial growth
C	No microbial growth	Microbial growth	Microbial growth
D	No microbial growth	No microbial growth	No microbial growth

Question 14

The graph below shows the outcomes of dialysis and transplant for patients with kidney failure.

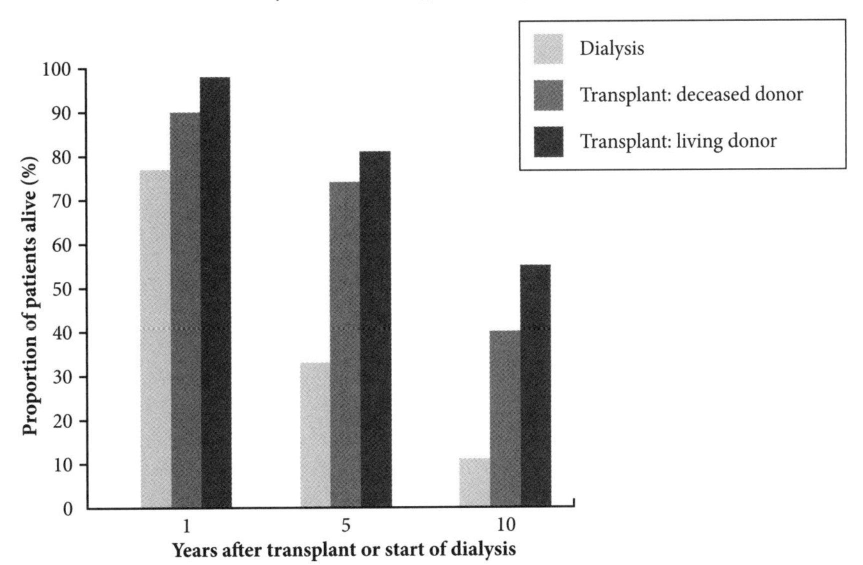

Which of the following statements is correct?

A Dialysis does not improve the life span of a person with kidney failure.

B It is more effective to have dialysis for 1 year than to have a transplant that lasts for 10 years.

C The chances of long-term survival are improved by receiving a transplant from either a living or a deceased donor.

D People with kidney failure should wait until a kidney from a living donor is available, rather than receive dialysis or a transplant from a deceased donor.

Question 15 ©NESA 2021 SI Q15

An example of the mutagenic effect of ultraviolet radiation (UV) on DNA is shown in the diagram.

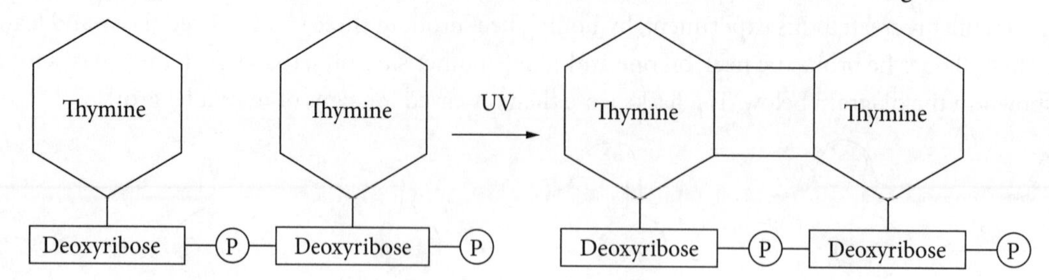

What is the mutagenic effect that is modelled?

A Thymine is duplicated.

B Bonds are formed between adjacent bases.

C Nucleotides form bonds in the backbone of DNA.

D Thymines on the two strands of DNA form bonds.

Question 16 ©NESA 2021 SI Q17

The images show the sequence of changes in the chromosomes (stained black) during mitosis in plant cells.

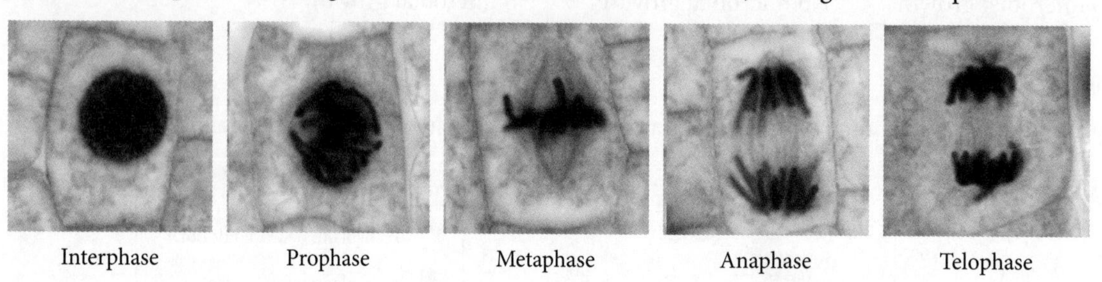

Interphase Prophase Metaphase Anaphase Telophase

Steve Gschmeissner / Science Photo Library

Which statement is true for mitosis?

A Crossing over occurs during prophase.

B Sister chromatids separate during anaphase.

C Two daughter cells are produced during telophase.

D Homologous pairs of chromosomes line up in metaphase.

Question 17 ©NESA 2011 SI Q17

The pedigree shows the inheritance of a trait controlled by a pair of alleles.

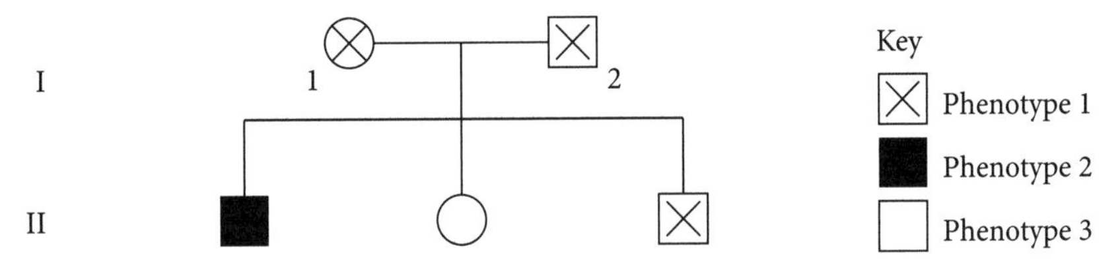

Which Punnett square correctly represents the cross between the parents in generation I?

A

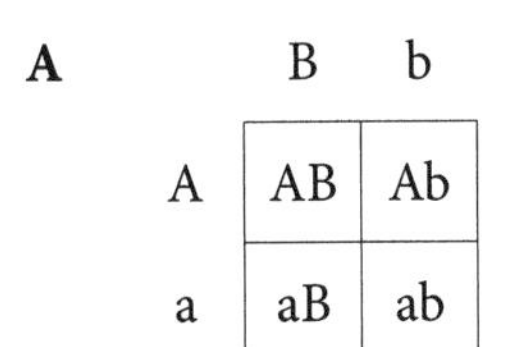

	B	b
A	AB	Ab
a	aB	ab

B

	B	B
A	AB	AB
A	AB	AB

C

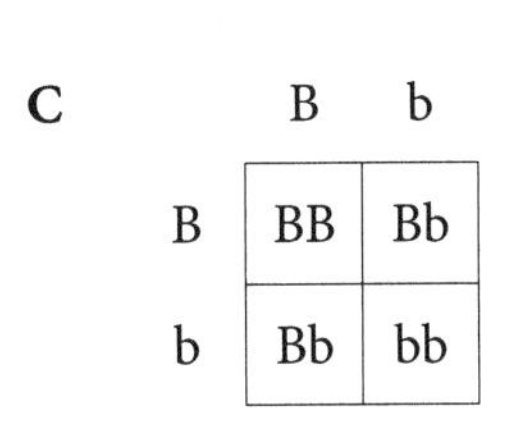

	B	b
B	BB	Bb
b	Bb	bb

D

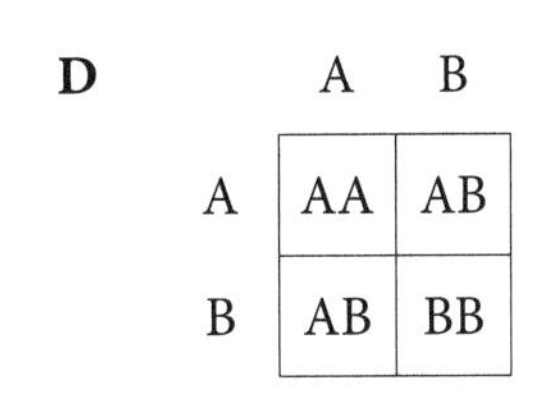

	A	B
A	AA	AB
B	AB	BB

Use the information below to answer Questions 18 and 19.

Melanomas are characterised by uncontrolled cell division caused by mutations that continue to occur once the tumour has developed. Scientists have discovered that vaccines produced using antigens extracted from the patient's own melanoma cells can be useful in treating melanoma. When injected, the vaccines stimulate an immune response.

Question 18 ©NESA 2016 SI Q19

What can be inferred from the scientists' discovery?

A Cancer cells carry unique antigens.

B Self-antigens are not present on cancer cells.

C The melanoma patient has a dysfunctional immune system.

D The body cannot mount an immune response against cancer cells.

Question 19 ©NESA 2016 SI Q20

The effect of the melanoma vaccine is to stimulate

A T cells which produce antibodies.

B cytotoxic T cells which activate B cells.

C cell division to produce more lymphocytes.

D production of B cells which destroy melanoma cells.

Question 20

Simple sequence length polymorphisms (SSLPs) are repeated sequences of bases located in the regions of DNA between genes. Variations in the length of SSLPs can be used to understand the genetic variation between two individuals.

The diagram below shows the three alleles of a single SSLP, and the DNA fingerprints that result from the six possible genotypes for this SSLP.

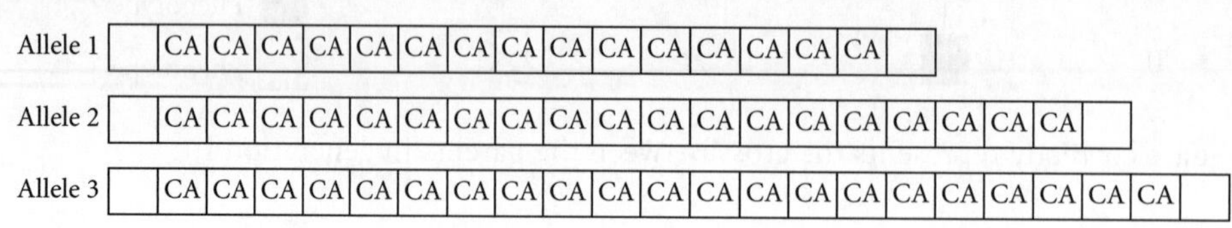

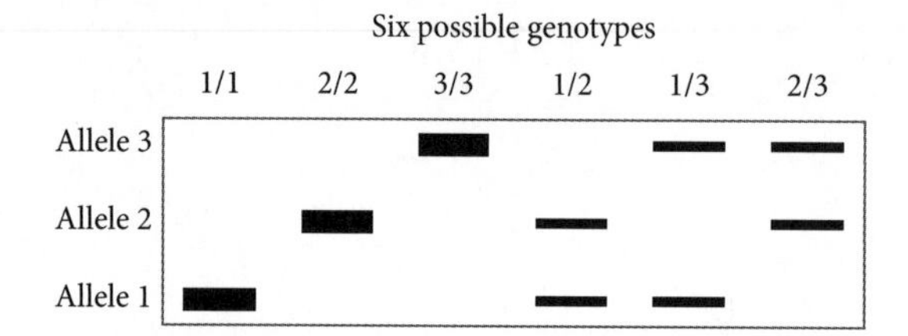

Two individuals, both with the genotype 2/3, have four children together.

Which of the DNA fingerprints below is the most likely result?

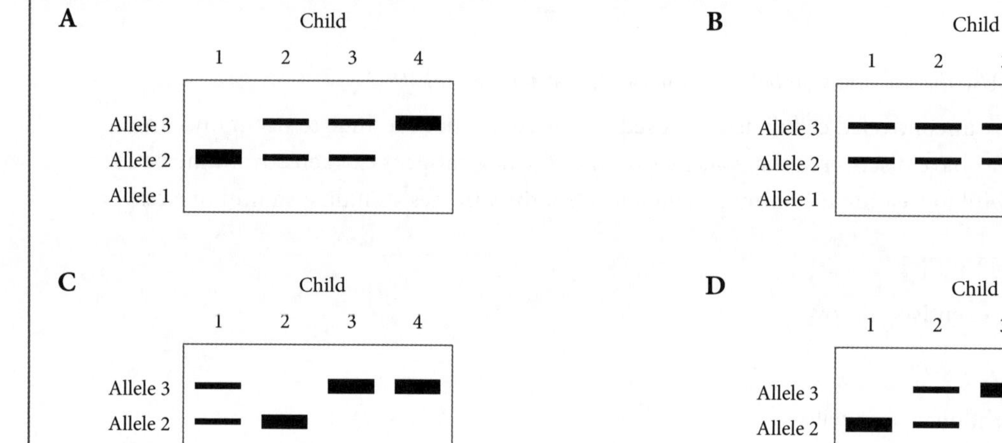

Section II

80 marks
Attempt Questions 21–31
Allow about 2 hours and 25 minutes for this section

Instructions

- Answer the questions in the spaces provided. These spaces provide guidance for the expected length of response.
- Show all relevant working in questions involving calculations.
- Extra writing space is provided at the back of this booklet. If you use this space, clearly indicate which question you are answering.

Question 21 (3 marks) 

The following diagram shows a mutation.

Original DNA sequence

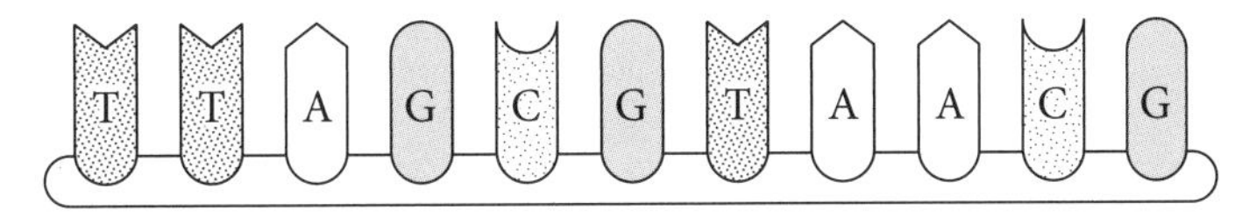

Mutated DNA sequence

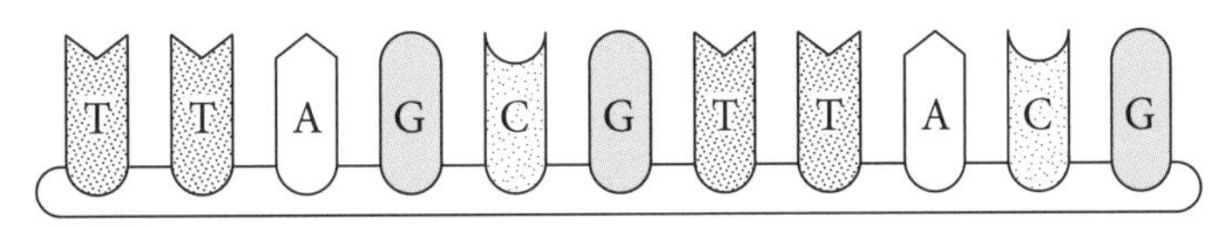

a What type of mutation is shown in the diagram? 1 mark

---

b Outline another type of mutation. 2 marks

End of Question 21

Question 22 (3 marks)

Marfan syndrome is an autosomal dominant condition that affects the heart, blood vessels, bones and eyes. It is caused by a mutation in the *FBN1* gene, which is usually inherited from a parent with the condition. However, in 25% of cases, it occurs in people without any family history, as a result of a new, or de novo, mutation. De novo mutations often result in 'mosaicism', where an individual has some cells with the normal *FBN1* gene and some cells with the mutated *FBN1* gene, as shown in the diagram below.

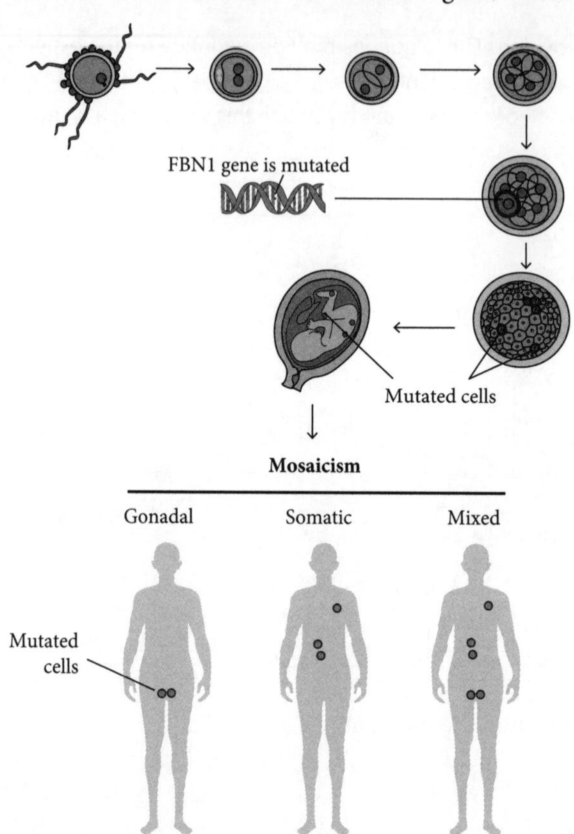

Complete the table below to outline the impact of each type of mosaicism on the individual in which the mutation occurs, as well as their potential offspring.

Type of mosaicism	Description	Impact on individual	Impact on individual's offspring
Gonadal	Mutation present in the germ-line cells of the gonads only		
Somatic	Mutation present in various organs but not the gonads		
Mixed	Mutation present in various organs as well as the gonads		

End of Question 22

Question 23 (7 marks)

Many pathogens are food-borne or water-borne.

a With the use of a diagram, outline how pathogens can pass from one host to another via contaminated food and water sources. 3 marks

b Design an experiment that could be used to test a food **or** water source for the presence of microbes. 4 marks

End of Question 23

Question 24 (4 marks)

Compare features of the innate and adaptive immune responses that protect a host against an invading pathogen.

Question 25 (9 marks)

Many people are exposed to noise levels that can damage their ears and lead to hearing loss. One profession that experiences higher than average noise levels is carpentry, where carpenters are exposed to sounds from saws, drills and other tools.

The graph below compares the average hearing levels of two 55 year olds, one of whom is a carpenter, showing how loud a sound must be at various frequencies for the person to be able to hear it.

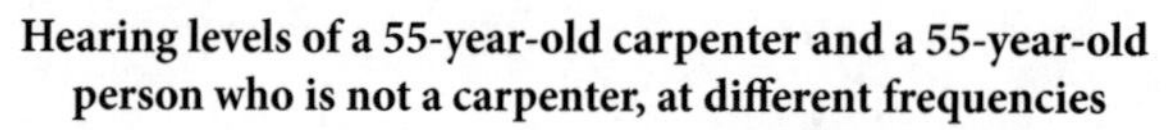

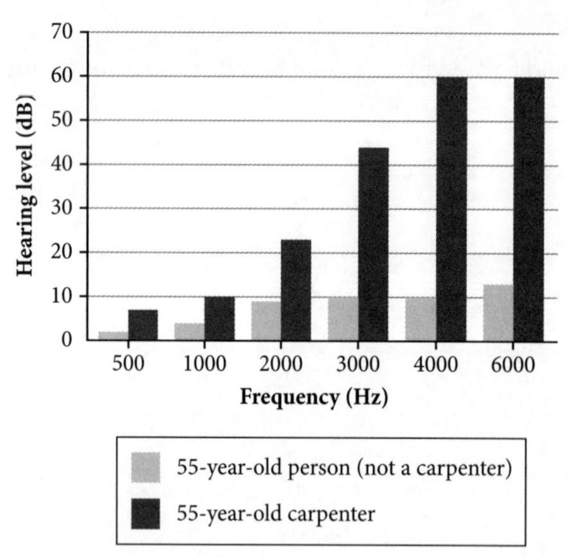

Question 25 continues on page 155

a Convert the data from the column graph into a line graph on the grid below. 4 marks

b Which of the two graphs is most useful in presenting the data? Justify your answer. 2 marks

c Hearing damage such as this occurs when the hairs in the cochlea are damaged. Justify the use of a particular technology that could be used to improve the hearing of a 55-year-old carpenter. 3 marks

End of Question 25

Question 26 (6 marks)

In the 1920s, the koala, *Phascolarctos cinereus*, was in danger of becoming extinct due to excessive hunting. In an effort to protect the species, 18 animals from Victoria were released into the Flinders Chase National Park on Kangaroo Island. In 2019, the Department for Environment and Water estimated the Kangaroo Island koala population to be 48 000.

During the summer of 2019–2020, unprecedented bushfires around Australia reduced the Kangaroo Island population to approximately 8500.

Assess the impact of mutation, gene flow and genetic drift on the gene pool of the Kangaroo Island koala population between 1920 and 2020.

End of Question 26

Question 27 (5 marks) ©NESA 2012 SII Q26

A scientist performed an epidemiological study to investigate the cause and effect relationship of smoking and lung cancer as follows:

1 Handed out a scientifically valid questionnaire to all colleagues (n = 144) at work

2 Checked that there were an equal number of male and female respondents

3 Discovered that there were more non-smoking respondents than smoking respondents. Removed some of the non-smokers until both groups had equal numbers

4 Checked that all the respondents had a medical check-up in the past year

5 Analysed data, wrote the paper and published it in a scientific blog.

From the information provided, analyse the methodology used by this scientist.

Question 28 (9 marks)

Throughout history, there have been numerous epidemics, with the most notorious being outbreaks of bubonic plague. This disease is now known to be caused by the bacterium *Yersinia pestis*, carried by rats, mice and other rodent species. It is spread to humans primarily through fleas that bite the rodents and then jump to a human host, but it can also spread when humans handle the tissue or blood of an infected animal, or inhale droplets in the air.

Symptoms include vomiting, nausea, fever, swollen lymph nodes and skin sores that turn black, leading to the common name, 'the Black Death'. Outbreaks of plague still occur today, but rapid diagnosis and treatment with antibiotics allow most people to survive, where once many people would have died.

The most infamous epidemic of the plague originated in China in 1334, spreading along the trade routes and reaching Europe via ports in Sicily by the late 1340s, killing approximately 25 million people, or one-third of the total population of Europe. It remained for centuries and spread to North America in the early 1900s by ship.

Question 28 continues on page 158

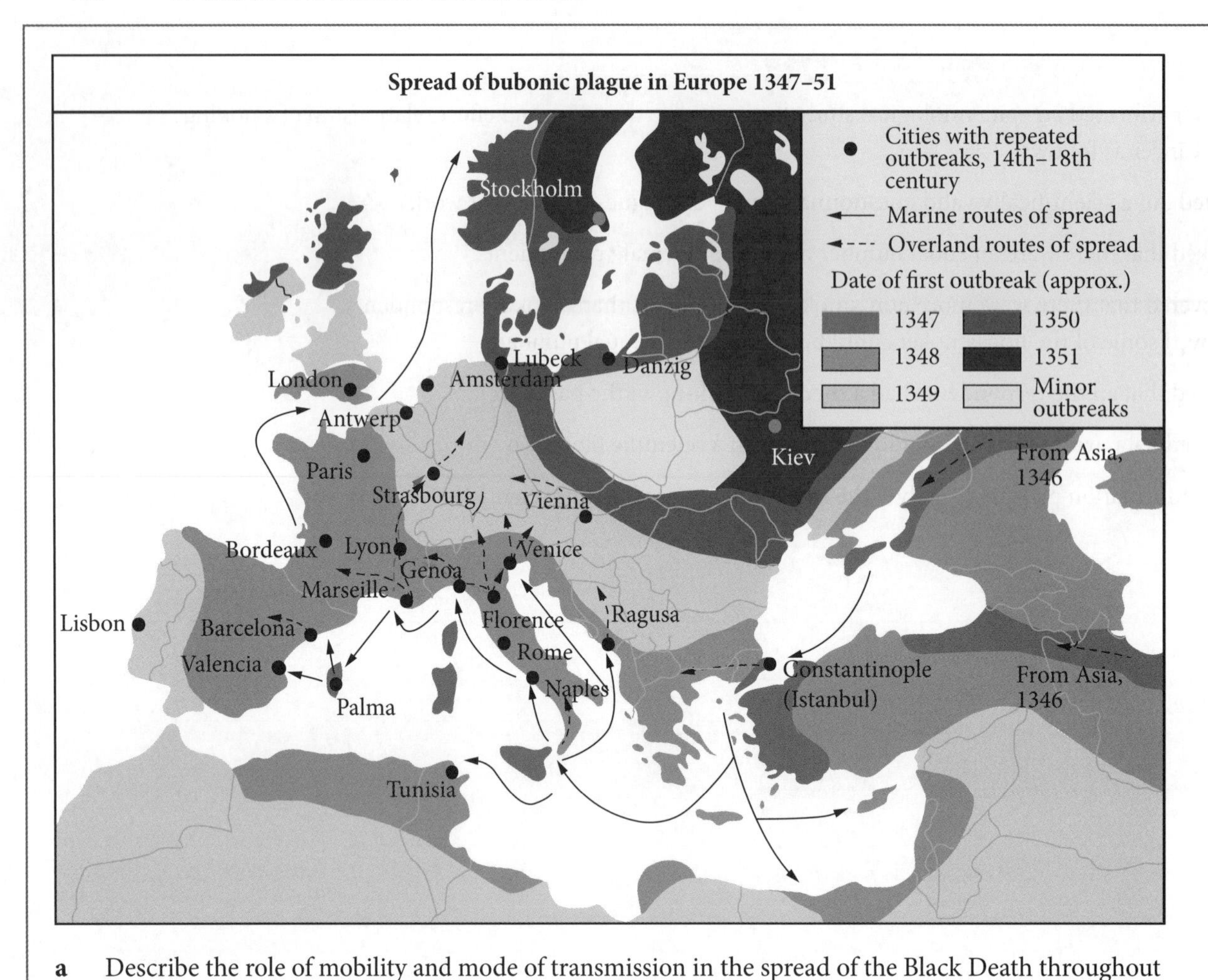

a Describe the role of mobility and mode of transmission in the spread of the Black Death throughout Asia and Europe. 4 marks

Question 28 continues on page 159

b In the 14th century, people had little understanding of the causes and transmission of disease. However, authorities understood that the disease could pass from one infected person to another, leading to some of the first public health measures, including the following.

- Medical inspections were conducted by 'plague doctors' to confirm suspected cases.
- Plague doctors wore long leather coats, gloves and a hat, with a beak-shaped mask filled with herbs, as shown in the diagram. This prevented them from breathing in the bad-smelling air associated with plague.
- Infected people and their families were isolated in their homes or in plague hospitals.
- Ships were restricted to port for 40 days (leading to the term 'quarantine', from *quaranta*, the Italian word for 'forty').
- The movement of people and goods was controlled.

history_docu_photo / Alamy Stock Photo

Assess the effectiveness of **two** of the measures undertaken by the authorities in the 14th century, and suggest additional measures that could have been implemented if they had had a modern understanding of disease transmission. 5 marks

End of Question 28

Question 29 (5 marks) ©NESA 2014 SII Q29 ●●●

Scientists have tried to achieve a viable embryo by fusing two ova (eggs) from the same female.

Explain whether the offspring produced using this process would be a clone of the female whose two ova were used. Use your knowledge of gamete formation and sexual reproduction to support your answer.

End of Question 29

Question 30 (9 marks)

Analyse this statement:

'As our understanding of disease transmission and the immune response to pathogens has increased, scientists have developed measures to prevent, treat and control the spread of disease. Antibiotics and vaccinations have prevented the deaths of millions of people since their development. However, both antibiotics and vaccines have the potential to lose their effectiveness over time.'

End of Question 30

Question 31 (20 marks)

Growth hormone (GH) is a small protein that stimulates growth. It is released by the anterior pituitary gland. Levels of GH are highest during childhood and adolescence, and progressively decrease with age.

GH is secreted in response to growth hormone-releasing hormone (GHRH), which is produced by the hypothalamus. GH causes the liver to increase its production of insulin-like growth factor 1 (IGF-1). IGF-1 stimulates the growth of muscles and bones and inhibits the production of GHRH by the hypothalamus.

a In the space below, construct a negative feedback loop to represent the control of GH levels outlined above. 3 marks

Question 31 continues on page 163

b Growth hormone deficiency (GHD), also called pituitary dwarfism, occurs when the pituitary gland cannot produce sufficient growth hormone. Children with GHD have abnormally short stature with normal body proportions. The condition is typically diagnosed around age 5, or at the onset of puberty.

Congenital GHD is present at birth. There are several types of congenital GHD, each caused by a different mutation:

- Type I is caused by an autosomal recessive mutation in the *GH1* gene.
- Type II is caused by an autosomal dominant mutation in the *GHRH* gene.
- Type III is caused by an X-linked recessive mutation in the *BTK* gene.

A woman with congenital GHD marries a man who does not have GHD, and has no family history of the disease. They have one daughter, who also has congenital GHD.

i ●●● What type of congenital GHD does the woman have? Justify your conclusion, using Punnett squares to support your response. 5 marks

Question 31 continues on page 164

ii Congenital GHD type I can be further divided, based on the level of GH produced:

- Type IA – no GH is produced.
- Type IB – a small amount of GH is produced, but levels are still well below normal.

Upon diagnosis of congenital GHD, children are treated with daily injections of GH.

Children with type IB typically respond well to GH treatment and, if treated early, will often grow to reach the typical height of their family members. Children with type IA typically respond well to GH treatment at first, but they soon develop antibodies in response to the hormone. As a result, they typically remain very short into adulthood.

Using your understanding of the specific immune response, explain the varying success of GH treatments in children with type IA and type IB congenital GHD. 4 marks

Question 31 continues on page 165

c The following passage is an extract from a journal article.

> Growth hormone (GH) was first isolated from the human pituitary gland in 1956. By 1960, clinical trials had shown that GH derived from the pituitary glands of human cadavers (dead bodies) was an effective treatment for GHD in children. Between 1963 and 1985, about 27 000 children worldwide received GH derived from human cadavers. Because it was in short supply, GH treatment was reserved for only the most severe cases of GHD.
>
> In 1985, the US Food and Drug Administration (FDA) received reports that four young adults had contracted Creutzfeldt-Jacob Disease (CJD), a rare and fatal prion disease, typically seen in older adults. Upon further investigation, researchers noticed that all four had been previously treated with GH. The use of cadaver-derived GH abruptly ceased, due to the risk of contamination and disease transmission.
>
> Discontinuation of cadaver-derived GH led to rapid approval of the first genetically engineered GH. It was first produced in 1981 by California biotech company Genentech, by inserting the gene that codes for GH into *E. coli* bacteria. The recombinant GH, called Protropin, was approved in 1985, just 6 months after the discontinuation of cadaver-derived GH. It is now the primary source of GH used to treat GHD. It is also used to treat other conditions, such as Turner syndrome and chronic kidney disease.
>
> Adapted from: Ayyar, V. (2011) History of growth hormone therapy. *Indian Journal of Endocrinology and Metabolism*, 15(3), 162–5. https://doi.org/10.4103/2230-8210.84852

Explain how developments in our understanding of genes and genetic technologies have led to improvements in the treatment and prevention of non-infectious diseases. Refer to the information provided throughout this question in your response. 8 marks

END OF PAPER

SECTION II EXTRA WORKING SPACE

Biology

PRACTICE HSC EXAM 2

General instructions

- Reading time: 5 minutes
- Working time: 3 hours
- Draw diagrams using pencil
- Calculators approved by NESA may be used

Total marks: 100

Section I – 20 marks

- Attempt Questions 1–20
- Allow about 35 minutes for this section

Section II – 80 marks

- Attempt Questions 21–30
- Allow about 2 hours and 25 minutes for this section

Section I

20 marks
Attempt Questions 1–20
Allow about 35 minutes for this section

Circle the correct multiple-choice option for Questions 1–20.

Question 1

Which of the following is true regarding internal fertilisation in nature?

A It does not require sexual intercourse.

B It occurs only in terrestrial organisms.

C It produces offspring that are genetically identical.

D It increases the likelihood of the successful union of gametes.

Question 2

The diagram below shows the mammalian eye.

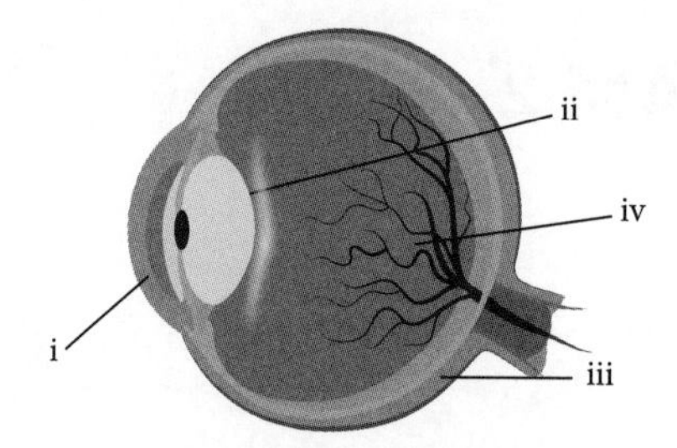

Which row in the table below correctly identifies a structure labelled in the diagram, and its function?

	Label	Structure	Function
A	i	Cornea	Refraction of incoming light rays
B	ii	Lens	Opens and closes to control the amount of light entering
C	iii	Conjunctiva	Protects the eye from minor mechanical damage
D	iv	Aqueous humour	Provides nutrition and maintains pressure in the eye

Question 3

Wheat is one of Australia's most important agricultural crops, worth more than $5 billion each year. Australia is free of many diseases that threaten the wheat industry, such as karnal bunt. Karnal bunt is caused by a fungus called *Tilletia indica*, which produces black, powdery spores that discolour the grain and give it a fishy smell. The disease originated in India and is now found in parts of North America, South America and Asia.

If this disease entered Australia through infected imported seeds, it could have a devastating effect on our economy, with other countries refusing to import Australian wheat.

Which control measure would be most effective at preventing karnal bunt from entering Australia?

A Random inspections of wheat crops in Australia should be performed, looking for signs of contamination.

B Farmers and importers of wheat to Australia should sign a declaration stating that their crops are free of karnal bunt.

C Authorities should inspect wheat seeds as they enter ports, keeping them quarantined for the entire incubation period.

D The importation of flour, bread and other baked goods containing wheat grown in countries with karnal bunt should be banned.

Refer to the following information to answer Questions 4, 5 and 6.

StarLink corn was a genetically modified corn that contained a gene from the common soil bacterium *Bacillus thuringiensis*. This gene allowed the corn to produce an insecticide protein called Cry9C. In 1998, StarLink corn was approved for use in animal feed; however, it was not approved for use in human food, because further studies were needed to determine whether Cry9C was safe for human consumption.

In 2000, traces of StarLink corn were found in products made for human consumption, including corn chips and taco shells. Consumers of the affected corn claimed that the Cry9C protein caused an allergic reaction. Although studies revealed that this was not the case, affected products were recalled and StarLink corn was removed from the market.

Adapted from Agricultural Biotechnology Council of Australia, 2012, 'StarLInk Corn', *ABCA Issues Paper* 6, vol. 2, https://www.abca.com.au/wp-content/uploads/2012/09/ABCA_IssuesPaper_6_v2.pdf

Question 4

What term best describes a plant such as StarLink corn, that has had its genome modified to include DNA from another species?

A Clone

B Hybrid

C Prokaryotic

D Transgenic

Question 5

Food allergies occur when the immune system reacts to foreign substances in food that are harmless for most people.

If Cry9C was found to cause allergies, what term could be used to describe it?

A Antibody

B Antigen

C Immunoglobulin

D Pathogen

Question 6

It was hypothesised that StarLink corn entered the human food chain when it was planted too close to unmodified crops.

Which of the following processes would account for this hypothesis?

A Artificial pollination

B Gene flow

C Genetic drift

D Mutation

Question 7

Hyperopia is a visual disorder, also known as longsightedness. People with hyperopia are able to see objects in the distance but struggle to see objects that are close to them.

What is a possible cause of hyperopia?

A The lens refracts incoming light rays too much, so the image falls behind the retina.

B The lens refracts incoming light rays too much, so the image falls in front of the retina.

C The lens does not refract incoming light rays enough, so the image falls behind the retina.

D The lens does not refract incoming light rays enough, so the image falls in front of the retina.

Use the following information to answer Questions 8 and 9.

Pathogens can often remain active outside the body of a host, before losing their ability to cause infection. Once expelled from a host's body, the pathogen may settle on a hard surface such as a table, door handle or tap, where it may be picked up by a new host.

The table below shows the average amount of time that some common pathogens remain infectious outside a host's body on hard surfaces.

Name of pathogen	Time remains infectious
Rhinovirus	24 hours
Respiratory syncytial virus	6 hours
Influenza	24 hours
Salmonella	4 hours
Norovirus	3–4 days
E. coli	24 hours

Question 8

What adaptation may these pathogens have that would increase their ability to remain infectious on surfaces outside a host?

A Low tolerance of environmental antimicrobial substances

B Heavy particles, so they remain suspended in the air

C Resistance to drying out and changes in oxygen concentration

D Surface proteins that allow adherence to host cell surface receptors

Question 9

Although many respiratory viruses are able to survive for long periods of time on hard surfaces, their ability to infect a new host is quickly reduced on skin surfaces such as the hands. The table below compares the survival time on hard surfaces with that on skin surfaces.

	Approximate time the pathogen remains infectious	
	Hard surfaces	**Skin surfaces**
Rhinovirus	24 hours	1 hour
Respiratory syncytial virus	6 hours	20 minutes
Influenza	24 hours	5–15 minutes

Why will pathogens not survive as long on skin surfaces as on hard surfaces?

A Pathogens are destroyed by antimicrobial chemicals on the skin.

B The pathogen cannot withstand the high temperatures of the body.

C It is not possible for pathogens to enter the host's body once they have adhered to the skin.

D Hygiene practices such as washing hands prevent pathogens from adhering to the surfaces.

Use the following information to answer Questions 10 and 11.

The diagram below shows a genetic technique that occurs *in vitro* (outside a living organism).

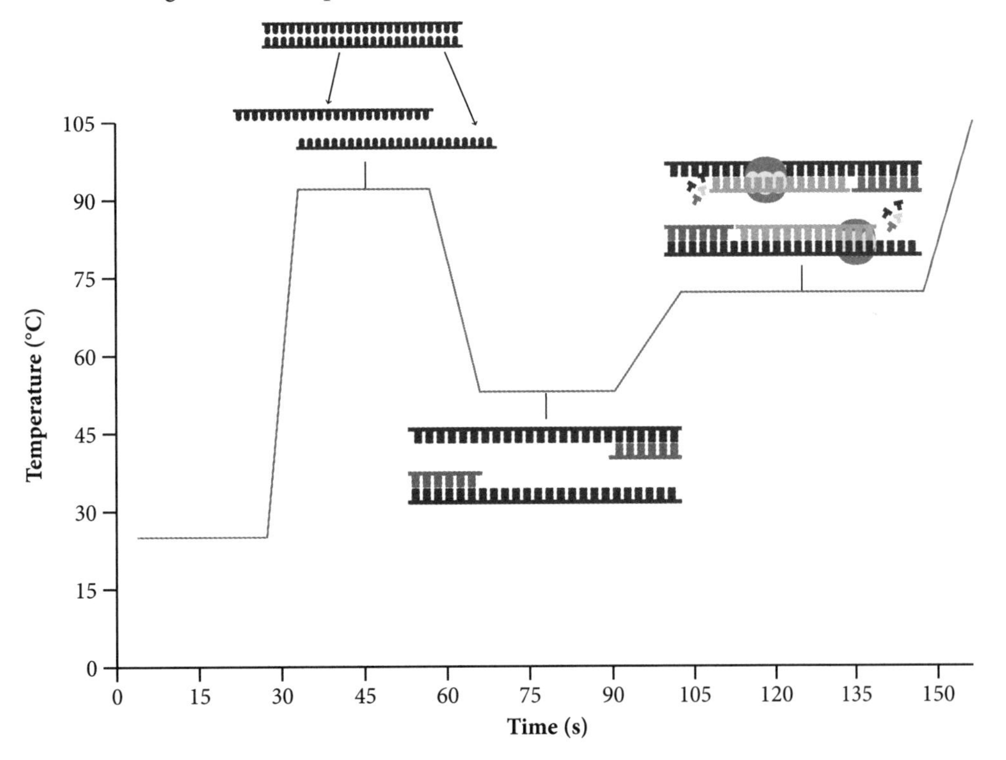

Question 10

What *in vivo* (within a living organism) process does this technique most closely resemble?

A Mitosis

B Fertilisation

C DNA replication

D Polypeptide synthesis

Question 11

Which of the following correctly identifies the process that occurs at 92°C?

A Annealing

B Denaturation

C Division

D Extension

Question 12

Diseases can be transmitted from one host to another via direct contact, indirect contact or a vector. The information below describes how a variety of diseases are transmitted.

- Ringworm is transmitted when an infected person's skin touches an uninfected person's skin.
- Glandular fever is transmitted through saliva, usually through kissing but sometimes through sharing utensils or water bottles.
- Hepatitis is transmitted through contaminated food or water sources.
- Zika virus is transmitted through the bite of a mosquito, and occasionally through sexual intercourse when body fluids are exchanged.
- Influenza is transmitted by respiratory droplets picked up from a contaminated surface.
- Malaria is transmitted by the bite of a mosquito.

Based on the information above, which row of the table correctly classifies the mode(s) of transmission for each disease?

	Indirect contact	Direct contact	Vector
A	Hepatitis Influenza Ringworm	Glandular fever Hepatitis Ringworm	Malaria Zika virus
B	Glandular fever Hepatitis Influenza	Glandular fever Ringworm Zika virus	Malaria Zika virus
C	Influenza Zika virus	Glandular fever Hepatitis	Malaria Ringworm
D	Glandular fever Zika virus	Hepatitis Influenza	Malaria Ringworm

Question 13 ©NESA 2021 SI Q14

The human *PRNP* gene codes for the prion protein. Misfolding of this protein may be the result of ingesting tissue that contains misfolded prion protein or a mutation. Accumulation of misfolded prion protein causes serious diseases such as Creutzfeldt–Jakob disease (CJD).

Which of the following statements best classifies CJD?

A It is both a genetic disease and an infectious disease.

B It is a genetic disease only, since it is encoded by a gene.

C It is not an infectious disease because the prion is non-cellular.

D It is an infectious disease, and the normal prion protein is the pathogen.

Question 14

The gene for coat colour in cats is carried on the X chromosome.

'O' represents the allele for orange fur, and 'B' represents the allele for black fur. Tortoiseshell cats exhibit fur of both colours in their coats, giving them a mottled or 'tortoiseshell' appearance.

Which of the following litters could be produced from a mating between an orange female cat and a black male cat?

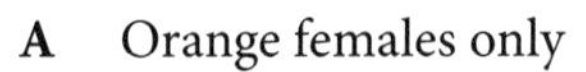

A Orange females only

B Tortoiseshell males and females

C Orange females and orange males

D Tortoiseshell females and orange males

Question 15

A student wrote the following statement in her notes, but her friend did not think it was correct.

'The immune system cannot destroy a pathogen once it has entered a host's cell.'

Which statement is correct?

A The student is correct. Antibodies can only deactivate a pathogen in the bloodstream or fluid outside cells.

B The student is incorrect. Antibodies are injected into the infected cells by memory B cells to destroy the infected cell.

C The student is correct. Cytotoxic (killer) T cells can only deactivate a pathogen in the bloodstream or fluid outside cells.

D The student is incorrect. Cytokines are injected into the infected cells by cytotoxic (killer) T cells to destroy the infected cell.

Question 16

In some low-income countries, young children often suffer from micronutrient deficiencies, which affect their growth and development.

The graph below shows the prevalence of some micronutrient deficiencies in Vietnamese children, from a 2012 study.

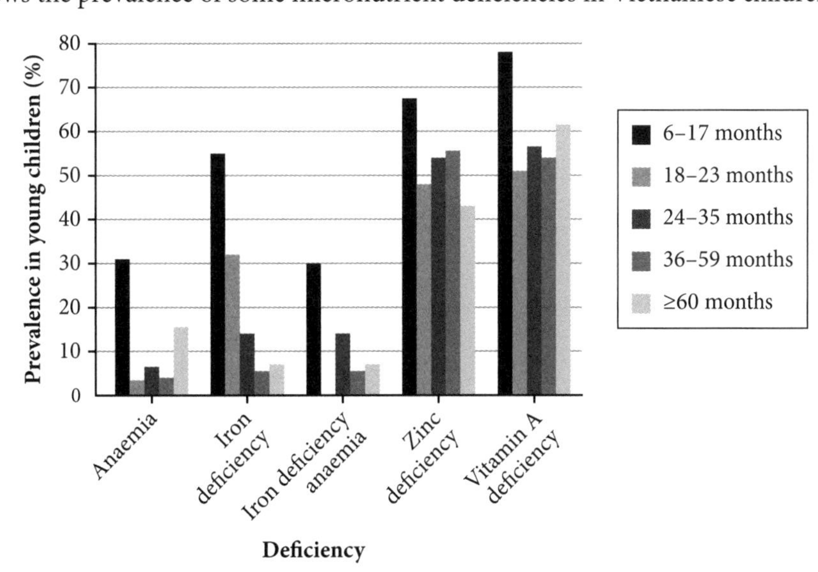

Which of the following statements is correct?

A Children aged 18–23 months cannot develop iron deficiency anaemia.

B Children under 6 months of age do not suffer from any micronutrient deficiencies.

C The prevalence of all micronutrient deficiencies increases as the age of the child increases.

D Vitamin A deficiency is the most prevalent micronutrient deficiency in the children studied.

Use the following information to answer Questions 17 and 18.

The desert locust *Schistocerca gregaria* is a species of short-horned grasshopper found in Africa, the Middle East and Asia. The desert locust feeds on grain, corn, cotton and vegetable crops. It is considered the most destructive migratory pest in the world; it can travel over 100 kilometres in a single day in search of food, and a swarm may destroy an entire crop in only a few days. Males have 23 chromosomes, while females have 24 chromosomes. This difference in chromosome number occurs because males have only one sex chromosome, while females have two.

Question 17

Which of the following is true regarding reproduction in *S. gregaria*?

A Male offspring are produced asexually.

B Sperm may contain 11 or 12 chromosomes.

C Only half the sperm produced by males are capable of fertilising an egg.

D Characteristics carried on the sex chromosomes are only inherited by females.

Question 18

The fungus *M. anisopliae* acts as a natural biopesticide that is effective against desert locusts. The fungus infects the locust and multiplies within its body, taking around 21 days to kill its host. Scientists are investigating how the fungus might be genetically modified to increase its virulence.

Which of the following is a potential benefit of using genetically modified *M. anisopliae* against desert locusts?

A It would reduce reliance on traditional pesticides.

B The GM fungus could be less cost-effective than traditional pesticides.

C The fungus could affect other insects, leading to a loss of biodiversity.

D The fungus requires specific environmental conditions to survive and multiply.

Question 19

Testicular cancer originates in the testes of a man. In 2021, there were 980 new cases of testicular cancer diagnosed in Australia. The population of Australia was approximately 25 million, with 49.8% of the population male and 50.2% female.

Which of the following statements about testicular cancer is correct?

A The incidence rate was 3.92 per 100 000.

B The incidence rate was 7.87 per 100 000.

C The prevalence rate was 3.92 per 100 000.

D The prevalence rate was 7.87 per 100 000.

Question 20

Translocation Down syndrome is the cause of about 4.5% of Down syndrome cases. In translocation Down syndrome, an extra copy of chromosome 21 is attached to chromosome 14. It is sometimes called familial Down syndrome, because the mutation is passed on from a parent via the sperm or egg to their offspring.

The parent is said to be a carrier, because they themselves do not have features of Down syndrome. This is because the parent typically has a 'balanced translocation'. That is, they have the correct amount of genetic material present, but in the wrong location (as shown in the diagram below).

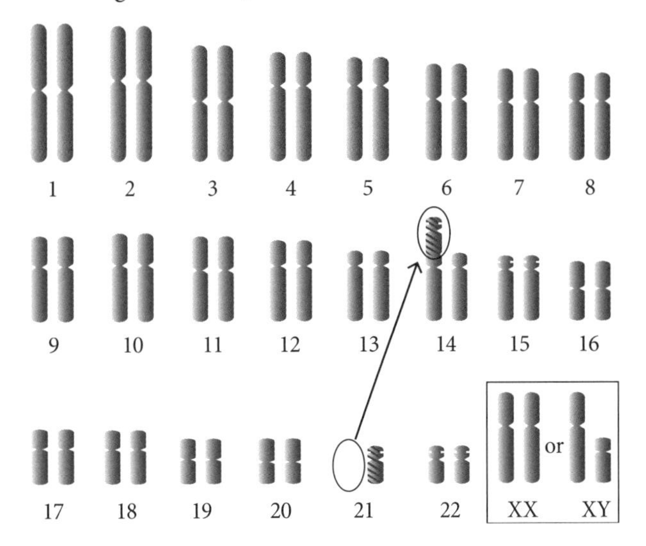

A man who is a carrier of translocation Down syndrome marries a woman who is genotypically normal. They seek the advice of a genetic counsellor, as they are considering trying to conceive.

Which of the following statements is correct?

A 50% of their children will be carriers of translocation Down syndrome.

B There is no chance that future offspring will inherit translocation Down syndrome.

C There is a 50% chance that a zygote they produce will contain only 45 chromosomes.

D There is a 50% chance that their offspring will have a copy of chromosome 21 attached to chromosome 14.

9780170465250

Section II

80 marks
Attempt Questions 21–30
Allow about 2 hours and 25 minutes for this section

Instructions

- Answer the questions in the spaces provided. These spaces provide guidance for the expected length of response.
- Show all relevant working in questions involving calculations.
- Extra writing space is provided at the back of this booklet. If you use this space, clearly indicate which question you are answering.

Question 21 (7 marks) 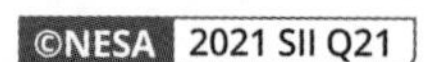 

a Label **two** features on the diagram below that would help to classify this pathogen as a bacterium. 2 marks

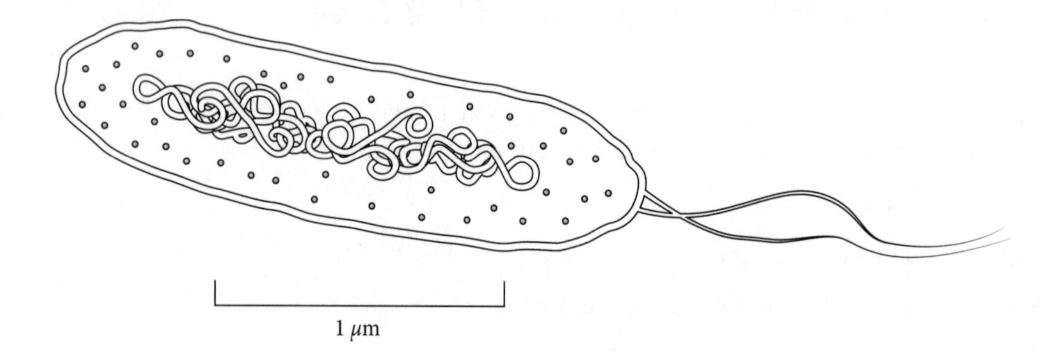

Question 21 continues on page 177

b A scientist followed Koch's postulates to confirm that this bacterium was causing diarrhoea in pigs on a local farm.

Complete the boxes in the flow chart provided to show the steps taken by the scientist. 2 marks

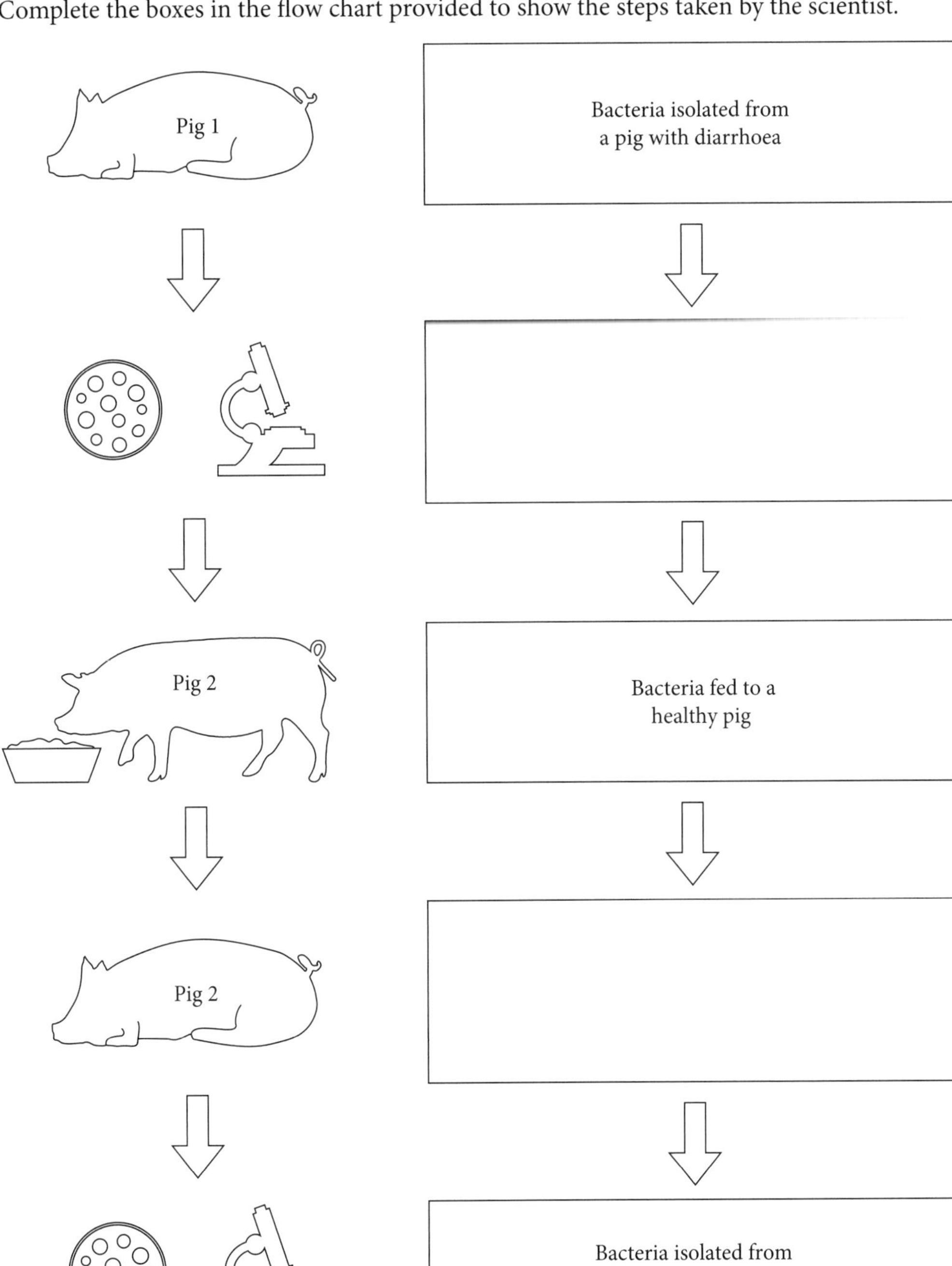

Question 21 continues on page 178

c Two pig farmers on neighbouring farms noticed that their pigs were suffering from diarrhoea and gradually losing weight. The farmers each adopted a different strategy to deal with this disease, as shown in the table.

Farm	Strategy	Result
1	Treatment with antibiotics	All pigs recovered after two weeks
2	Elimination of rats and mice from pig sheds to improve hygiene	Decrease in number of sick animals over three months

Outline **one** benefit and **one** limitation of the strategies used on each farm. 3 marks

Question 22 (3 marks)

The diagram below shows part of the process of polypeptide synthesis.

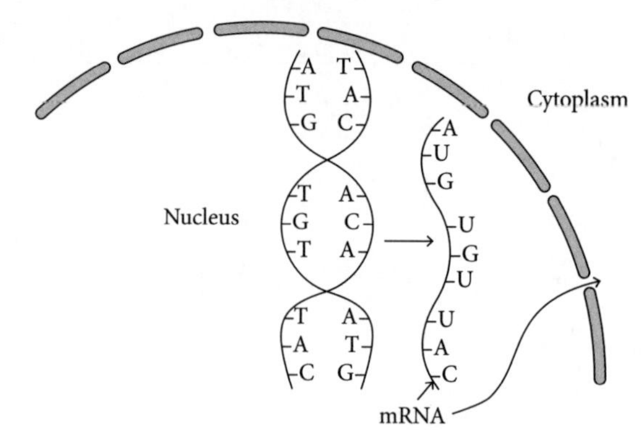

a Name the process that produces mRNA. 1 mark

b Explain why mRNA follows the path shown in the diagram. 2 marks

End of Question 22

Question 23 (4 marks)

The diagram below shows a model of DNA structure and replication.

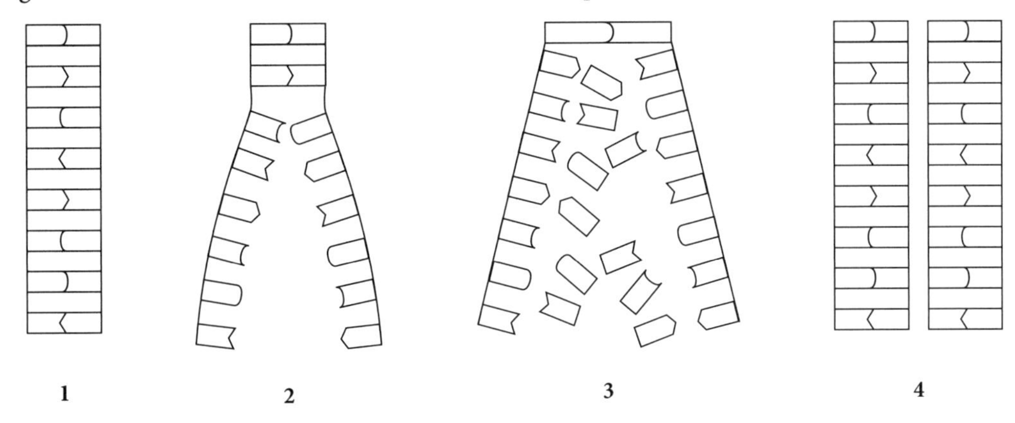

Complete the table below.

	Modelling of DNA structure	Modelling of DNA replication
Outline one aspect that is accurate		
Outline one aspect that is inaccurate		

Question 24 (9 marks)

Gestational diabetes is a condition that many women develop during pregnancy, whereby their body does not produce enough insulin to regulate their blood glucose levels at the normal level of approximately $90\,mg\,dL^{-1}$. It is most common in women who are over 25, are overweight or obese, have a family history of diabetes or have other medical conditions that affect their hormones.

To test for gestational diabetes, pregnant women take a glucose tolerance test between 24 and 28 weeks of pregnancy. The woman fasts for at least 8 hours, then her starting blood glucose is taken from a blood sample. The woman then consumes a sugary fluid, and her blood glucose levels are monitored over the next 2 hours.

The graph below shows the results of the glucose tolerance test for three women: a non-diabetic, a diabetic who has received no treatment and a diabetic who is treated with insulin.

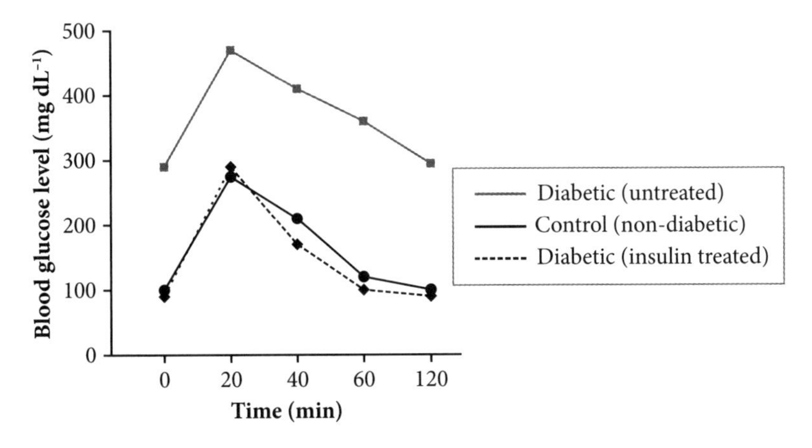

Question 24 continues on page 180

a Analyse the trends shown in this graph. 3 marks

b In the space below, construct a negative feedback loop to show what would be happening in the non-diabetic woman following her consumption of the sugary fluid. 3 marks

c The rate of gestational diabetes has increased in recent years. Describe features of an effective educational campaign that could be used to prevent cases of gestational diabetes. 3 marks

End of Question 24

Question 25 (12 marks)

The fruit of the oil palm, *Elaeis guineensis*, provides about 35% of the world's vegetable oil. Modern *E. guineensis* has three fruit forms: *dura* (thick-shelled), *pisifera* (shell-less) and *tenera* (thin-shelled). Shell thickness is determined by a single gene, called the *Shell* gene.

The diagram below shows the inheritance of shell thickness in *E. guineensis*. The parental generation plants are homozygous.

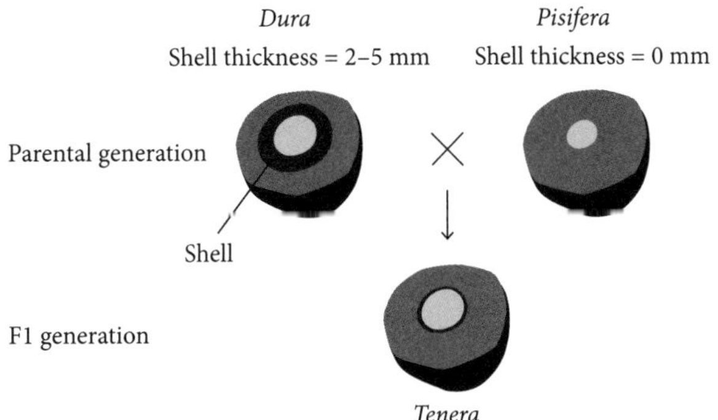

Use the following symbols for alleles in your answers.

Allele	Symbol
Thick shell	T
No shell	N

a What is the genotype of the *dura* plant in the parental generation? 1 mark

b The *tenera* form is the basis of commercial palm oil production, as it yields far more oil than the *dura* and *pisifera* forms.

A farmer wishing to enter the palm oil industry crossed two *tenera* plants and was disappointed to find that only some of the offspring were of the *tenera* form.

Explain why this happened. Include a Punnett square to support your explanation. 4 marks

Question 25 continues on page 182

9780170465250

c In order to mass-produce palms with the *tenera* form, commercial palm oil famers use the tissue culture technique shown below.

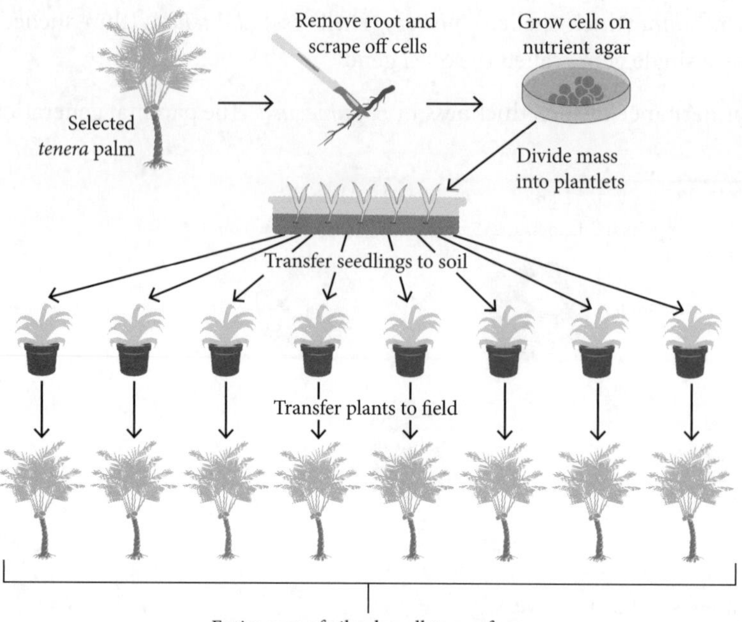

i Explain why the method shown would always produce palms of the *tenera* form. 2 marks

ii Although *E. guineensis* is originally native to West Africa, 85% of the global palm oil supply comes from South-East Asia. In this region, palm oil plantations are the main driver of destruction and clearing of tropical rainforests. For example, on the island of Borneo, at least 50% of all deforestation between 2005 and 2015 was related to oil palm development.

Using the information provided in this question, assess the impact of palm oil agriculture on biodiversity. 5 marks

End of Question 25

Question 26 (4 marks) ©NESA 2021 SII Q29

The koala is a mammal that maintains a stable body temperature of close to 36.6°C.

A study was conducted. Koalas were observed in natural forests in south-eastern Australia. Their posture in the tree and the ambient temperature were recorded. Ambient temperatures were divided into two categories, hot and mild.

The graph shows the posture of koalas observed.

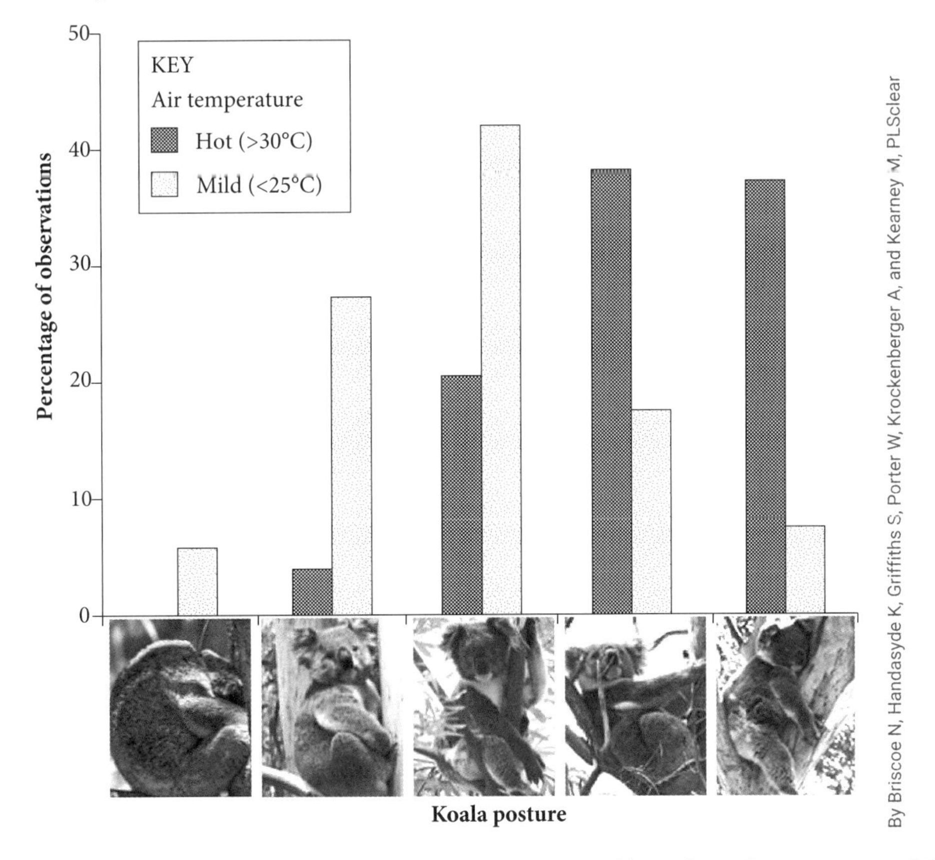

Additional data showed that the temperature of some tree trunks and large branches was up to 9°C cooler than air temperature during hot conditions.

Explain the adaptations used by the koalas in this study to maintain a stable body temperature. Make reference to the stimulus provided. 4 marks

Question 26 continues on page 184

Question 27 (9 marks)

The diagram below shows a DNA fingerprint and the apparatus used to produce it.

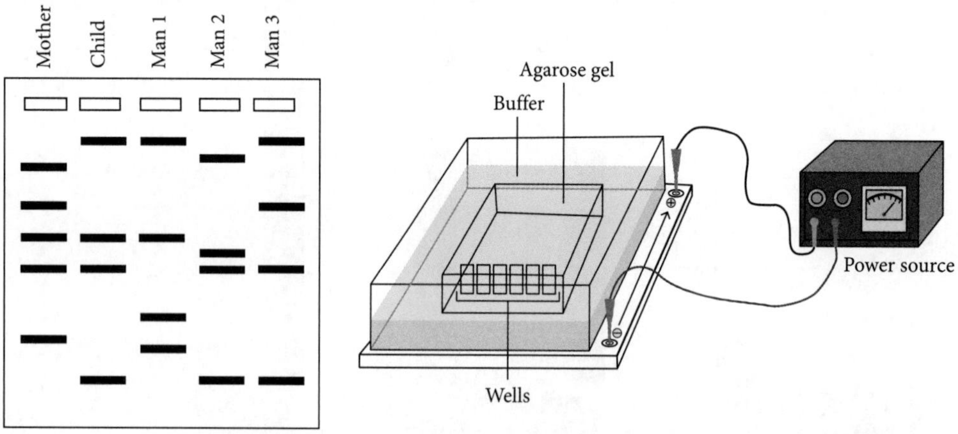

a Explain why scientists use a number of different restriction enzymes to produce a unique DNA fingerprint. 4 marks

b Describe how the apparatus above is used to produce the DNA fingerprint shown. 3 marks

Question 27 continues on page 185

c Outline another use of DNA fingerprinting technology besides that which is shown on page 184. 2 marks

Question 28 (4 marks) ©NESA 2021 SII, Q26

Zebra populations are suffering from a reduction in their gene pools due to habitat destruction and increasing isolation. This has led to an increase in the number of offspring born with coat patterns different to that of their parents. An example is shown.

Philip J. Briggs / Alamy Stock Photo

Explain possible reasons for the increase in these offspring. 4 marks

End of Question 28

Question 29 (12 marks)

Polycystic kidney disease (PKD) is a genetic disorder in which clusters of cysts develop primarily within the kidneys, causing the kidneys to enlarge and lose function over time. This can sometimes lead to kidney failure and the need for dialysis or a kidney transplant.

a PKD is an autosomal trait. What does this mean? 1 mark

b There are two main types of PKD:

- autosomal dominant PKD (ADPKD), which occurs in approximately 1 in every 400–1000 people
- autosomal recessive PKD (ARPKD), which occurs in approximately 1 in 20 000 people.

The pedigree below shows the inheritance of PKD in a family. Affected individuals are shaded.

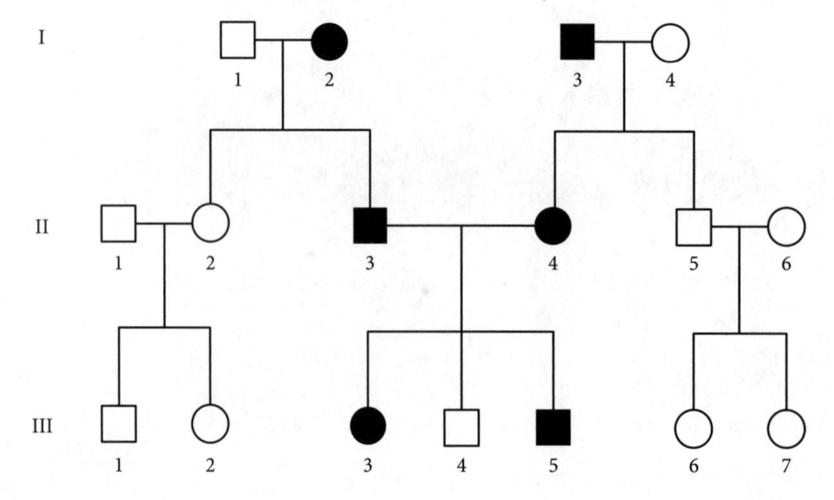

i Is PKD in this family dominant or recessive? Justify your choice with reference to specific individuals. 3 marks

Question 29 continues on page 187

ii ●●● About 30% of patients with ARPKD will have kidney failure by age 10, and about 50% of patients with ADPKD will have kidney failure by age 60. People with kidney failure require renal dialysis or a kidney transplant.

The following passage is an extract from a journal article.

> Patients with kidney failure face a shortage of kidneys from deceased human donors. As a result, patients spend an average of 3.9 years on dialysis while they wait for a suitable donor kidney. During this time, approximately 35% of patients die or develop comorbidities that make them unsuitable for a transplant.
>
> Domestic pigs could provide kidneys suitable for use in humans; they are anatomically and physiologically similar to humans and pig breeding is well practised and cost-effective. However, genetic differences between pigs and humans mean the transplanted organ is likely to trigger an immune response in the recipient. This would lead to failure of the transplanted organ.
>
> In an effort to overcome this rejection, scientists have used genetic engineering to 'knock out' the genes for the pig surface antigens. When used in conjunction with immunosuppressive medication, they have resulted in excellent kidney function in monkeys without any features of organ rejection. However, in the absence of innate and adaptive immune responses, the monkeys in each experiment succumbed to infection and died, despite having functioning kidneys.
>
> At present, the FDA recommends that a patient should only be considered for a pig organ transplant if their life expectancy is less than two years. It may be difficult to predict the exact survival of a patient, and many patients with a life expectancy less of than two years have comorbidities that make them unsuitable transplant candidates.
>
> Adapted from: Cooper D., Hara H., Iwase H. et al. (2020) Clinical pig kidney xenotransplantation: How close are we? *Journal of the American Society of Nephrology*, 31(1), 12–21, https://doi.org/10.1681/ASN.2019070651

Using the information provided, discuss the ethical and societal implications of pig organ transplantation. 8 marks

Question 29 continues on page 188

Question 30 (16 marks)

Cervical cancer occurs in the cervix, the part of the female reproductive system that connects the uterus to the vagina. Almost all cases of cervical cancer are caused by infection with a virus called human papilloma virus (HPV), which causes mutations in the cells of the cervix, resulting in cancer cells. HPV is transmitted through close contact with genital skin during sexual activity.

Female reproductive system

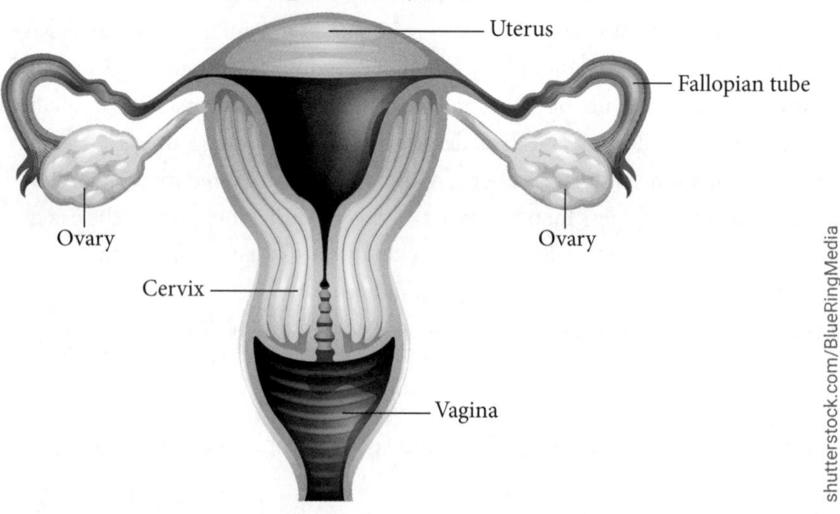

a There are many different mutations that may lead to changes to cervical cells. An example is shown below.

Original sequence	GAA	ATC	ACT	GAG	CAG	GAG	AAA
Mutated sequence	GAA	ATC	ACT	AAG	CAG	GAG	AAA

What type of mutation is shown in the example above? 1 mark

b Mutations occur in the cells of the cervix. Explain whether an affected person would pass this mutation on to their offspring. 3 marks

Question 30 continues on page 189

c Explain whether cervical cancer should be classified as an infectious or non-infectious disease. 3 marks

d Australia has one of the lowest incidences and mortality rates of cervical cancer in the world. This is largely attributed to an extensive screening program introduced in 1991 to detect abnormalities in cervical cells while pre-cancerous or in the early stages of the disease. This test was changed in 2017 so that rather than testing for cancerous or pre-cancerous cells, it tests for the presence of HPV. There are usually no symptoms with HPV infection and cervical cancer does not usually cause symptoms until it is advanced, so early detection of the disease is vital.

The introduction of a vaccine against HPV in 2007 to all girls aged 12–13 (with a 'catch-up' program for those aged 14–26) and extended in 2013 to include boys aged 12–13, has led to further reductions and it is predicted that the disease may be eliminated in the future. Some experts believe that widespread screening of the population will no longer be necessary.

In every population there is a small proportion of people who are unable to be vaccinated, such as young children, those with allergies to the ingredients used to make vaccinations, and those receiving treatment that suppresses the immune response.

The graphs below show the projected changes to the incidence and mortality rates for cervical cancer due to vaccination, with and without a continued cervical screening program.

Incidence and mortality rate (age-standardised) of cervical cancer with vaccination, with and without cervical screening program

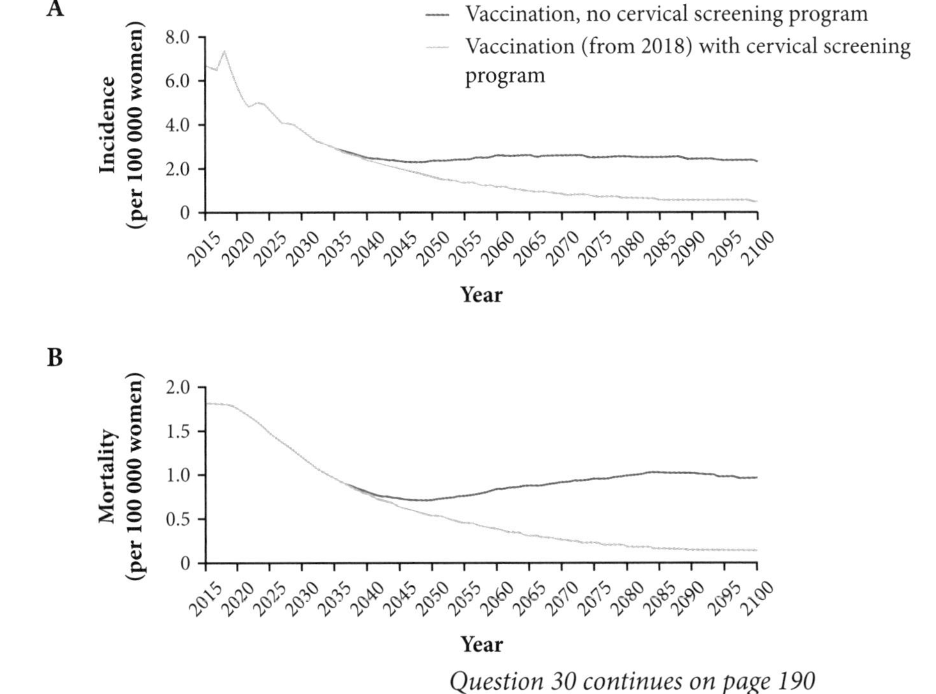

https://doi.org/10.1016/S2468-2667(18)30183-X

Question 30 continues on page 190

With reference to the information and graphs provided, evaluate the need for continued vaccination and screening programs for cervical cancer. 9 marks

END OF PAPER

SECTION II EXTRA WORKING SPACE

SECTION II EXTRA WORKING SPACE

SOLUTIONS

Test 1: Reproduction

Multiple-choice solutions

Question 1

C Binary fission

Binary fission is a form of asexual reproduction in which an organism grows and copies its genetic material, and then divides into two. **A** is incorrect because spores do not result from the splitting of the organism into two halves. **B** is incorrect because budding occurs when a small outgrowth on the parent organism breaks off to form a new organism. **D** is incorrect because sexual reproduction involves gametes, which are not present here.

Question 2

D fertilisation as a result of fusion of male and female gametes.

Sexual reproduction in plants involves gametes, which fuse together in the process of fertilisation. **A** is incorrect because although pollination is a form of sexual reproduction, it is not caused by the dispersal of seeds; it is caused by the dispersal of pollen. **B** is incorrect because cloning in plants is a type of asexual, not sexual, reproduction. **C** is incorrect because pollen grains are produced by meiosis, not mitosis.

Question 3

B II

Fertilisation occurs when a sperm fertilises an egg. In humans, this occurs in the fallopian tube, as shown at point II. **A** is incorrect because point I shows the egg being swept into the fallopian tube prior to fertilisation. **C** is incorrect because point III shows the first mitotic division, which occurs after fertilisation. **D** is incorrect because point IV shows implantation of the blastocyst.

Question 4

A II and III

All modes of sexual reproduction involve gametes (sex cells), which are produced by the process of meiosis. **B** is incorrect because spores are a feature of asexual reproduction. **C** and **D** are incorrect because sexual reproduction in most eukaryotic organisms does not involve implantation; it only occurs in some mammals.

Question 5

C Implantation occurred about week 4.

The developing embryo begins to secrete HCG shortly before it implants into the wall of the uterus. The increase in HCG at about week 4 shows that this is when implantation occurs. **A** is incorrect because oestrogen levels drop after birth; the graph shows the levels of oestrogen are high at 40 weeks, so birth could not have occurred at week 36. **B** is incorrect because fertilisation cannot occur at day 0, as ovulation is yet to occur – ovulation takes place approximately two weeks after the end of the last menstrual cycle. **D** is incorrect because the formation of the placenta begins at week 4 and is fully formed by week 20.

Question 6

C It involves fertilisation.

Sexual reproduction always involves fertilisation, the union of male and female gametes. **A** is incorrect because in many organisms, fertilisation occurs externally, and copulation (sexual intercourse) is not required. **B** is incorrect because in some organisms, such as plants, a single parent can produce both male and female gametes, which can undergo self-fertilisation. **D** is incorrect because identical twins are genetically identical, despite being the result of sexual reproduction.

Question 7

D follicle-stimulating hormone.

Follicle-stimulating hormone (FSH) causes follicle growth in the ovaries, eventually leading to an egg being released; with higher FSH levels, or ovaries that are more sensitive to FSH, multiple eggs may be released at the same time, leading to an increased likelihood of fraternal twins. **A**, **B** and **C** are incorrect because these hormones do not play a major role in stimulating follicle development; oestrogen causes thickening of the uterine lining, progesterone maintains the uterine lining, and luteinising hormone triggers ovulation.

Question 8

B An injection of an oxytocin receptor antagonist (OTR-A) is given, blocking the action of oxytocin.

According to the flow chart, oxytocin triggers uterine contractions and stretching of the cervix in a positive feedback loop that causes labour to progress. An injection that blocks the action of oxytocin would stop this positive feedback mechanism and would therefore slow labour. It would not be used to induce labour, as it would have the opposite effect. **A** and **C** are incorrect because both of these result in the cervix dilating, which, according to the flow chart, leads to oxytocin release, contractions and the onset of labour. **D** is incorrect because according to the flow chart, if the foetus drops lower in the uterus, the cervix will dilate, triggering the onset of labour.

Question 9

D Multicellular haploid

In plants, the haploid spores can develop by mitosis to produce a multicellular haploid plant. In humans, a haploid sperm or egg cannot undergo mitosis to produce a multicellular haploid organism. **A** is incorrect because the zygote stage occurs in humans after fertilisation takes place. **B** is incorrect because gametes are produced in humans by the process of meiosis. **C** is incorrect because humans are multicellular diploid organisms.

Question 10

D In diploid individuals, a second set of chromosomes may mask harmful effects of mutations in the first set.

Diploid organisms have two copies of each chromosome. If a harmful, recessive mutation occurs on one chromosome, the other normal chromosome would mask the mutation. This would be an evolutionary advantage of being diploid, because a recessive mutation on one chromosome would not adversely affect the organism. **A** is incorrect because spores are part of the haploid stage, so the fact that they are hardy would favour a haploid-dominant life cycle. **B** is incorrect because the fact that the diploid generation grows slowly and takes longer to reach reproductive maturity provides an advantage for the haploid generation, and would lead to the evolution of a haploid-dominant life cycle. **C** is incorrect because spores are part of the haploid stage, so the fact that they can travel long distances would lead to haploid offspring colonising the new environment; this would lead to the evolution of a haploid-dominant life cycle.

Short-answer solutions

Question 11

a The offspring would be clones of the parent (1 mark), meaning traits of the parent that make the strawberries suitable for farming and selling (e.g. sweet fruit, pest resistance) would be also found in the offspring (1 mark).

b The anthers are removed to prevent self-pollination (1 mark). This ensures that only the pollen from the plant of a different variety fertilises the ovules. This also ensures that the offspring are hybrids (1 mark).

Question 12

a Budding (1 mark)

b One of: fungi, protists, plants (1 mark)

c **i** Cross-fertilisation, self-fertilisation, asexual reproduction (1 mark)

ii Asexual reproduction is fast and produces offspring that are clones of the parent (1 mark). This would allow for the species to continue when the environment is favourable, through the very quick production of many well-suited offspring (1 mark). Sexual reproduction is slower but generates variety, especially if cross-fertilisation occurs (1 mark). If an environment changes or becomes less favourable, this variety allows for the continuity of the species because it is more likely that some individuals will be able to survive in the new conditions (1 mark).

Question 13

a Ovulation (1 mark)

b In non-pregnant women, progesterone (P4) levels peak around day 21 before decreasing back to day 0 levels, while in pregnant women they continue to increase beyond day 21 (1 mark). This is because the corpus luteum, which secretes P4, degenerates if pregnancy does not occur. If pregnancy does occur, the corpus luteum is maintained and P4 levels continue to rise (1 mark). hCG production begins about day 21 in pregnant women but is not present in non-pregnant women (1 mark). This is because hCG is produced by the cells surrounding the fertilised egg after it implants in the uterus – therefore it is only present if pregnancy occurs (1 mark).

For both P4 and hCG, you must state the difference between the two graphs (2 marks), and provide a reason for the difference (2 marks). The question is asking about differences, so it is not necessary to mention LH, as the levels of LH are the same in both graphs.

Question 14 ©NESA 2020 MARKING GUIDELINES SII Q25 (ADAPTED)

a There is little difference between the mean numbers of young/eggs produced in animals using the two modes of reproduction (1 mark), and the variability of the data is large, as shown by the standard deviations (1 mark). Therefore, no conclusion can be drawn regarding the effect of internal versus external fertilisation on the number of young/eggs produced (1 mark).

In order to score full marks, you need to identify that the data was inconclusive (or similar) regarding the effect of fertilisation type on the number of young/eggs produced. You must refer to the data (i.e. standard deviation and mean) to support your answer.

b The students have selected only a very few species for their study and either by chance or by specific selection of these species the number of eggs/young born are similar (1 mark). A much larger number of species should be included if the current hypothesis is to be reinvestigated (1 mark).

c Animals using external fertilisation will expend less energy on gestation (1 mark), as this will occur outside the body (1 mark).

Question 15

Agriculture involves the cultivation and breeding of animals, plants and other organisms to produce products that are useful to humans (1 mark). To maximise productivity and efficiency of milk production, farmers must have some scientific knowledge of reproductive processes in cattle.

To carry out artificial insemination, farmers must understand the oestrous cycle of female cattle (1 mark). If a farmer does not understand the signs of oestrus, the female may be inseminated at a time when there is low likelihood of pregnancy. It would probably take longer for a female to become pregnant and this would reduce the milk productivity of that female (1 mark). Therefore, understanding the oestrous cycle of female cattle has a positive impact (1 mark) on the ability to manipulate reproduction in agriculture through artificial insemination.

To ensure offspring have favourable characteristics, farmers must understand how characteristics such as sex, milk quality and volume are inherited (1 mark). If a farmer is aware of how sex is inherited, they can choose to artificially inseminate females with X-bearing sperm. This would ensure all offspring are females and can contribute to milk production (1 mark). Thus, scientific knowledge of how desirable characteristics are inherited has a positive impact on the manipulation of reproduction in agriculture by producing a greater proportion of favourable offspring (1 mark).

Scientific knowledge regarding the manipulation of cattle reproduction has a significant positive impact on agriculture. It allows farmers in the dairy industry to maximise the time that a female animal can be productive, as well as ensuring that the offspring produced have desirable characteristics.

For full marks, your answer must show an understanding of the term 'agriculture'. You must also identify a minimum of **two** areas of scientific knowledge. For each area, you need to describe why this knowledge is important and evaluate its impact on the manipulation of dairy cattle reproduction.

The areas of knowledge in the suggested answer are knowledge of oestrous cycles and inheritance patterns. Other areas of knowledge include how hormone levels vary during pregnancy, the factors influencing semen quality, the process of insemination, the role of hormones in inducing oestrus, and ultrasound as a method of confirming a successful pregnancy. The question requires you to apply your knowledge. To score full marks, you must relate your understanding of reproduction and productive technologies to the information provided in the stimulus material.

Test 2: Cell replication

Multiple-choice solutions

Question 1

A Uracil

Uracil is found in RNA, not DNA. **B** is incorrect because DNA contains four nitrogenous bases: adenine, thymine, cytosine and guanine. **C** and **D** are incorrect because each nucleotide in DNA contains a phosphate group and a deoxyribose sugar.

Question 2

C 8

A nucleotide consists of a single sugar, attached to a phosphate group and a nitrogenous base. **A, B** and **C** are incorrect because there are 8 nucleotides shown in the diagram.

Question 3

A Mitosis produces only two daughter cells, while meiosis produces four.

During mitosis, two daughter cells are produced. In meiosis, four daughter cells are produced. **B** is incorrect because the parent cell in both mitosis (the somatic cell) and meiosis (the germ-line cell) is a diploid cell. **C** is incorrect because in mitosis, the two daughter cells are genetically identical to each other and to the parent cell, while in meiosis, the four daughter cells are haploid, and are genetically unique. **D** is incorrect because mitosis consists of only one division, while meiosis consists of two divisions.

Question 4

B Meiosis

The diagram shows a pair of homologous chromosomes, each consisting of two sister chromatids, being separated. This only occurs during anaphase I in meiosis. Recombination has also occurred, which only occurs in meiosis. **A** is incorrect because homologous chromosomes do not line up in tetrads along the equator in mitosis. **C** is incorrect because cytokinesis is the process of dividing the cytoplasm, not dividing the contents of the nucleus. **D** is incorrect because DNA replication occurs before the step shown in the diagram.

Question 5

C somatic cell from a male human.

The karyotype shows 46 chromosomes, so it must be a somatic (body) cell from a human. It contains one X chromosome and one Y chromosome, so it must be from a male. **A** and **B** are incorrect because human gametes are haploid and contain only 23 chromosomes. **D** is incorrect because female somatic cells contain two X chromosomes.

Question 6

D One pair of homologous chromosomes

Two chromosomes are shown, each consisting of two identical sister chromatids. The chromosomes are a homologous pair, because they are the same length, have the same centromere position, and contain the same genes, indicated by the common banding pattern. **A** is incorrect because homologous chromosomes are not identical. Each one is inherited from a different parent, so although the genes on the chromosomes are the same, they do not necessarily have the same alleles. **B** is incorrect because a joined pair of sister chromatids are considered to be one chromosome. **C** is incorrect because there are two pairs of sister chromatids present.

Question 7

B

To produce a second generation, the complementary strands of each DNA molecule in the first generation separate. This results in a total of four template strands, two of which are from the original parent DNA. Each of the four strands serves as a template for a new strand, as shown in **B**. **A** is incorrect because there are four original parent DNA strands, where there should only be two. **C** is incorrect, because the two original parent DNA strands are paired; this is impossible as they would have separated and each served as a template for a new strand in the first round of replication. **D** is incorrect because it shows a combination of new and original DNA on the same strand, which does not occur.

Question 8

B DNA is always a double-stranded molecule.

This evidence supports the idea that DNA is double stranded, with A pairing with T, and C pairing with G. For example, if A accounted for 10% of bases, thymine would also account for 10%. The remaining 80% would be 40% C and 40% G. Hence, A (10%) and G (40%) add up to 50%. **A** is incorrect because DNA could be non-helical and still have 50% A and G bases, provided it was double stranded. **C** is incorrect because adenine does not pair with guanine; it pairs with thymine. **D** is incorrect because although the amount of A is always equal to the amount of T, and the amount of C is always equal to the amount of G, these four amounts are not always the same as one another (as shown in the example above).

Question 9

B S

At the start of the S phase, the cell contains one pair of chromosomes. After the S phase is complete, each chromosome consists of a pair of sister chromatids. These pairs are formed when the DNA in the chromosome is replicated. This means that DNA must have replicated during the S phase. **A** is incorrect because at the end of the G1 phase, the chromosomes do not consist of pairs of sister chromatids. This means DNA in each chromosome has not replicated during the G1 phase. **C** is incorrect because each chromosome already consists of pairs of sister chromatids at the start of the G2 phase. This means that DNA replication had already occurred prior to the start of the G2 phase. **D** is incorrect because each chromosome already consists of pairs of sister chromatids at the start of the M phase, and therefore replication must have occurred prior to the M phase. Pairs of sister chromatids are split during the M phase.

Question 10

C It allows organisms to grow and survive to reach reproductive maturity.

The cell cycle applies to mitosis, the process by which multicellular species grow and survive to reproductive maturity; reproductive maturity must be reached so that offspring can be produced to replace those that die, thereby ensuring continuity of the species. **A** is incorrect because the cell cycle only applies to mitosis, not meiosis. Meiosis is used for sexual reproduction in multicellular species, not mitosis. **B** is incorrect because the question is about multicellular species; prokaryotes are unicellular. **D** is incorrect because recombination only occurs in meiosis, not mitosis.

Short-answer solutions

Question 11

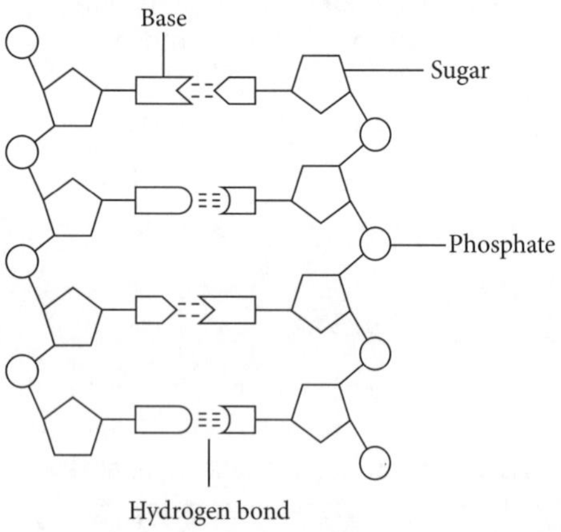

There are no half marks awarded in the HSC. All labels must be correct for 2 marks. At least one correct label would score 1 mark.

Question 12

a DNA replication (1 mark)

b
1 DNA helicase unzips the two DNA strands. (1 mark)

2 DNA polymerase adds complementary nucleotides according to the base pairing rule (A–T, C–G). (1 mark)

3 One strand (the leading strand) is synthesised continuously (1 mark) in the direction of unzipping.

4 The other strand (the lagging strand) is synthesised in sections. (1 mark)

5 Sections on the lagging strand are linked together by DNA ligase. (1 mark)

The question asks you to 'describe', not just outline. This means you must include a level of detail, such as the name of the enzyme involved, the base pairing rule, and the difference between the leading and lagging strands. You do not need to refer to RNA primers or RNA primase to score full marks, as this is not shown in the diagram.

Question 13

a E, A, C, B, D (1 mark)

b Any two of the following (1 mark each):

A – DNA condenses into visible chromosomes.
B – Spindle fibres separate sister chromatids and pull them towards opposite poles.
C – Chromosomes align along the equator of the cell.
D – Chromosomes have reached the poles; cytokinesis is occurring.
E – DNA is in the form of chromatin.

c Mitosis produces two identical daughter cells from a single parent cell (1 mark). In order to maintain the same number of chromosomes and the same genetic instructions (1 mark), the DNA must be copied before it is divided between the daughter cells.

> You must show that you understand that the daughter cells produced from mitosis should be identical to the parent cell, and therefore the DNA must be copied before the parent cell splits.

Question 14

a Any two of: chromosomes condense, nuclear membrane breaks down, chromosomes arrange in tetrads, spindle fibres attach to centromeres. (2 marks)

b (3 marks)

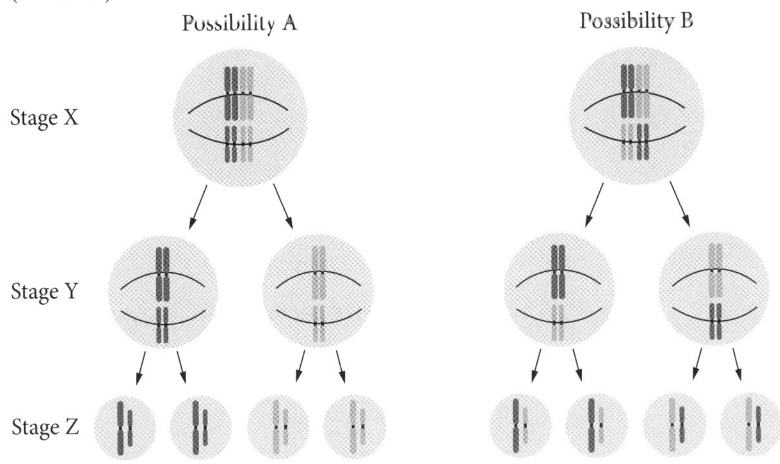

Question 15 ©NESA 2017 MARKING GUIDELINES SII Q24aii (DIAGRAM)

a **i** Aa Bb Gg (1 mark)

ii (1 mark)

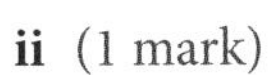

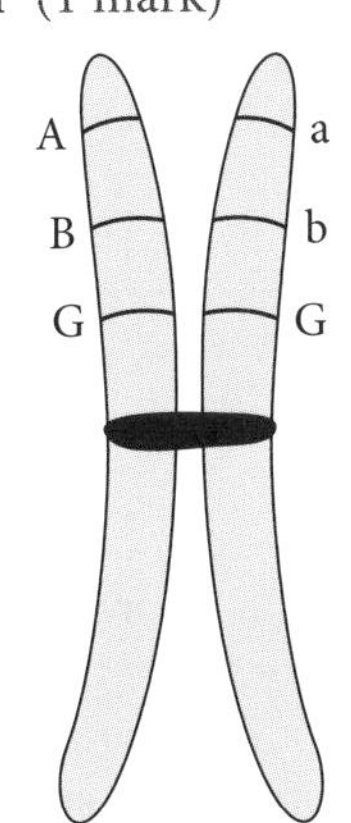

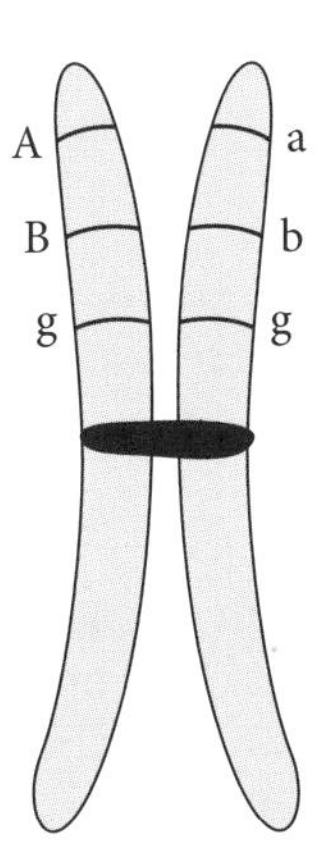

b Crossing over occurs when homologous chromosomes exchange sections during meiosis (1 mark). This means that some chromosomes will have both maternal and paternal alleles (1 mark). This increases the amount of variation in the offspring (1 mark).

Question 16

Random segregation results in gametes containing half the number of chromosomes of the parent cell (1 mark). This is because pairs of homologous chromosomes in the parent cell must segregate or split during meiosis (1 mark). Independent assortment results in gametes containing different combinations of maternal and paternal chromosomes (1 mark). This is because the alignment of one pair of homologous chromosomes along the equator during meiosis does not influence alignment of other chromosome pairs (1 mark). When two gametes combine during fertilisation, the full number of chromosomes is restored, producing a unique offspring with a unique set of genetic information (1 mark).

Variation in offspring is required for continuity of the species; if there is a change in the environment, large amounts of variation make it more likely that some individuals within the population will have traits that are suited to the selective pressure, and the species will continue to survive (1 mark). Thus, independent assortment and random segregation both help to ensure the continuity of the species, because they lead to variety in the haploid gametes produced by meiosis and, therefore, variety in the offspring (1 mark).

In your answer, you need to show a clear understanding of the processes of random segregation and independent assortment and how they lead to variation. You also need to identify the importance of variation in the continuity of species, and link this to the importance of the two processes.

Test 3: DNA and polypeptide synthesis

Multiple-choice solutions

Question 1

C It usually exists in the form of chromatin.

DNA exists in the form of chromatin for most of the cell cycle. **A** is incorrect because DNA in eukaryotic cells is also found in mitochondria and chloroplasts. **B** is incorrect because DNA is wrapped around histones, not contained within them. **D** is incorrect because chromatin only condenses into chromosomes during mitosis and meiosis.

Question 2

A W

A gene is a section of DNA that codes for the production of a polypeptide. **B** is incorrect because X depicts a single nucleotide, and genes are made up of multiple nucleotides. **C** is incorrect because Y depicts mRNA, which is single stranded, unlike genes, which are made of double-stranded DNA. **D** is incorrect because Z represents a polypeptide, which is a sequence of amino acids that is assembled according to the instructions within a gene.

Question 3

D the environment affecting phenotypic expression.

Environmental conditions can cause certain genes to be switched on (expressed) or switched off. Changes in gene expression may lead to changes in phenotype, in this case the plant's height and leaf size and shape. **A** is incorrect, as a cutting from the original parent plant would be a clone of the parent; there would be no genetic variation between the two plants. **B** is incorrect because, while it is possible that somatic mutations occurred during the growth and development of the cutting, it is unlikely that a somatic mutation would have such a profound impact on phenotype. **C** is incorrect because evolution does not occur at the individual organism level; it occurs at the population or species level.

Question 4

A the linear sequence of amino acids.

A protein's primary structure relates to which amino acids are in which position within a polypeptide chain. **B** and **C** are incorrect because the active, functional form of the protein exists when the polypeptide folds into three-dimensional structures. This only occurs at the secondary, tertiary and quaternary levels of protein structure. **D** is incorrect because primary structure relates to polypeptide structure, not mRNA structure.

Question 5

D polypeptides.

Ribosomes are the organelles that translate mRNA into sequences of amino acids (polypeptides). Therefore, if the function of ribosomes was impaired, they could no longer produce polypeptides. **A** and **B** are incorrect because, while tRNA and mRNA do interact with the ribosome during translation, ribosomes do not produce tRNA or mRNA; these are made in the nucleus in the process of transcription. **C** is incorrect because the ribosome does not produce amino acids; it merely links them together in the process of translation.

Question 6

C Tertiary

Tertiary structure relates to the overall three-dimensional structure of a single polypeptide chain. **A** and **B** are incorrect because, while human serum albumin does possess these two levels of structure, they are not the highest possible level. **D** is incorrect because quaternary structure only occurs in proteins that are made up of more than one polypeptide.

Question 7

B They lack ribosomes.

In order to synthesise polypeptides, ribosomes must translate a sequence of nucleotides in an mRNA molecule into a sequence of amino acids. **A** is incorrect because a gene is a section of DNA that codes for a polypeptide; if viral host cells can use viral DNA to produce polypeptides, then the DNA must contain genes. **C** is incorrect because all DNA is made of nucleotides. **D** is incorrect because, while it is true that viruses do not have a nucleus, this does not prevent polypeptide synthesis; for example, bacteria and other prokaryotes lack a nucleus to house their genetic material but can still carry out polypeptide synthesis.

Question 8

B CAU

The coding sequence of DNA is the same as the mRNA produced from it (except that U replaces T). The tRNA anticodon is complementary to an mRNA codon. If a coding sequence of DNA is GTA, the mRNA codon would be GUA, and the complementary anticodon would be CAU. **A** and **B** are incorrect because RNA does not contain thymine. **D** is incorrect because a GUA of tRNA would correspond to a CAT triplet of coding DNA.

Question 9

D For Trait *B*, genes have a greater effect on phenotype than the family environment does.

For Trait *B*, the similarity of identical twins is 90%, while the similarity of adoptive siblings is only 20%. This indicates that genes (which identical twins share) have a much stronger influence over the trait than the shared family environment does. **A** is incorrect because mutations that occur after the embryo has split may cause the twins to have genetic differences. **B** is incorrect because the percentage difference between identical twins and adoptive siblings is larger for Trait *B* than Trait *A*. This indicates that genes have a stronger influence over Trait *B*. **C** is incorrect because neither genes nor family environment appear to have strong influence over Trait *C*. This can be seen from the low percentage of similarity across all three types of siblings.

Question 10

C It allows for many different polypeptides to be produced from a single gene.

Splicing out selected exons allows one gene to code for multiple different polypeptides, which will each have a different sequence of amino acids. **A** is incorrect because introns are removed after transcription from the mRNA, not from the DNA. **B** is incorrect because exons are the amino acid coding sections; in some cases, exons are removed by splicing, but some exons are always left behind to form the mature mRNA. **D** is incorrect because by removing exons in some instances, splicing allows two cells that express the same gene to produce different polypeptides.

Short-answer solutions

Question 11 (3 marks)

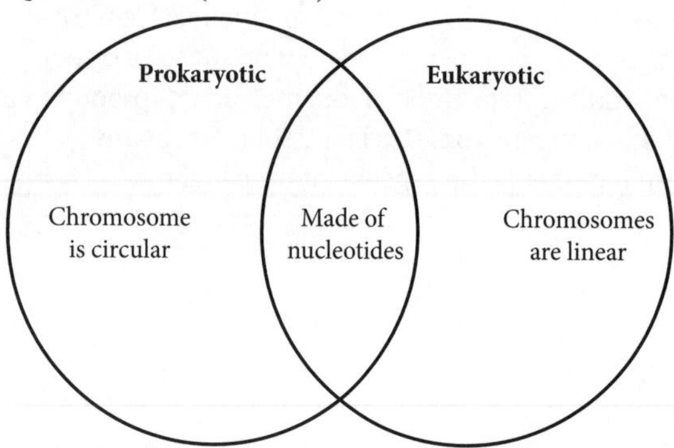

1 mark for each correct section of the Venn diagram. There are a number of possible correct answers. The similarities and differences must relate to features of DNA (e.g. DNA structure, DNA location), and not features of eukaryotic and prokaryotic cells in general.

Question 12

DNA is a double helix, consisting of two strands of nucleotides joined by complementary base pairs. RNA consists of only a single strand of nucleotides. (1 mark)

The nucleotides in DNA contain four different nitrogenous bases: adenine, thymine, guanine and cytosine. In RNA nucleotides, uracil replaces thymine. (1 mark)

Question 13

Polypeptides are made up of a series of amino acids joined by peptide bonds. (1 mark)

Proteins are made of one or more of these polypeptides (1 mark), folded into a three-dimensional shape. (1 mark)

Question 14

a RNA polymerase (1 mark)

b (3 marks)

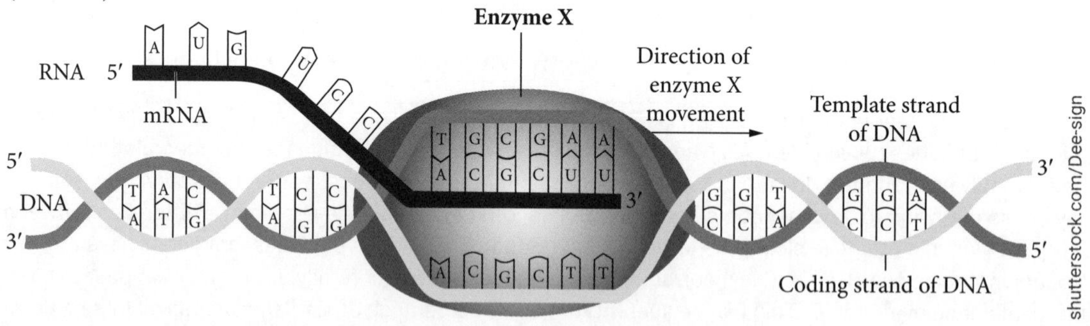

Diagram must show (1 mark each):

- RNA nucleotides attach to the template strand (not the coding strand).
- RNA is produced in the correct direction, according to the direction of RNA polymerase movement (i.e. the mRNA should be extending outside the transcription bubble on the left-hand side, not the right-hand side, because the 5' end would have been synthesised first).
- The base sequence in the mRNA is correct, including U replacing T.

c The mRNA base sequence is complementary to that of the DNA template strand (1 mark). The mRNA base sequence is the same as the DNA coding strand (1 mark), except that U replaces T (1 mark).

You should use the term 'complementary' rather than general terms such as 'matching'.

Question 15

a Ribosome (1 mark)

b mRNA is a single strand of nucleotides with the nitrogenous bases exposed (1 mark). The exposed bases of the mRNA allow the message to be 'read' by the ribosome in sets of three, called 'codons' (1 mark).

tRNA is a strand of nucleotides that fold into a 't' shape. Three exposed bases (an anticodon) on one end of the tRNA determine which amino acid is carried on the opposite end (1 mark). The exposed anticodon allows tRNA to bind to the complementary codon on the mRNA, bringing with it the corresponding amino acid, which is added to the growing polypeptide (1 mark).

c Glycine (or Gly) (1 mark)

Question 16

A gene is a section of DNA that codes for the production of a polypeptide. DNA is a double-stranded molecule made of nucleotides, each of which contains a sugar, a phosphate and a base (adenine, thymine, cytosine and guanine) (1 mark).

A polypeptide is a chain of amino acids linked by peptide bonds. Proteins are made of one or more polypeptide chains, folded into a three-dimensional structure. Proteins carry out many vital functions. For example, enzymes are globular proteins that catalyse metabolic reactions within the cell, while fibrous proteins play an important role in maintaining cell structure (1 mark).

Genes have a profound effect on proteins because they provide the 'instructions' for a protein's structure. During polypeptide synthesis, a gene is transcribed or copied into a molecule of mRNA and carried to a ribosome. The ribosome 'reads' the mRNA in sets of three bases, called codons, and translates it into a polypeptide. This polypeptide folds to form a protein (1 mark).

The sequence and type of amino acids in a polypeptide is determined by the sequence of DNA bases in the gene. If the nucleotide sequence of a gene changes, the sequence of amino acids may also change, and this can have an impact on the shape and chemical properties of the protein. For example, a change in a gene could change the shape of an enzyme's active site, rendering it unable to catalyse chemical reactions (1 mark).

Proteins also influence genes by controlling gene expression. DNA wraps around proteins called histones, to form chromatin. Altering the strength of the interactions between DNA and histones can switch genes 'on' or 'off' (1 mark). In this way, proteins control the expression of genes.

Enzymes such as DNA polymerase and DNA ligase interact directly with genes during the process of DNA replication, copying genes prior to the process of cell division. The action of these enzymes will determine whether the code within the genes is correctly copied. DNA repair enzymes can recognise and correct physical damage in DNA and prevent permanent changes to the genetic code (1 mark).

While proteins have an influence over the copying and expression of genes, genes have a more powerful influence over proteins because they provide the 'instructions' for a protein's structure (1 mark).

You must describe the structure and function of both genes and proteins. The verb is 'evaluate', so you must make a judgement about the effect of genes on proteins, and the effect of proteins on genes. You must explain the effects in detail; merely identifying or listing effects is not sufficient to score full marks – the mechanism behind the effect must be explored.

Test 4: Genetic variation

Multiple-choice solutions

Question 1

B Homozygous

Individuals that have two identical alleles for a particular gene are referred to as 'homozygous' for that gene. **A** and **C** are incorrect because both these terms refer to individuals that have two different alleles for a particular gene. **D** is incorrect because the genotype 'pp' would result in white flowers. A 'P' allele must be present in purple-flowered plants.

Question 2

B 2

When gametes are formed, pairs of alleles segregate, so each gamete contains one allele for each gene. **A** is incorrect because one possible gamete would only exist for an individual who is homozygous for both genes (e.g. AABB). **C** is incorrect because there are only two (not three) possible combinations: AB or aB. **D** is incorrect because four possible gametes would only exist for an individual who was heterozygous for both genes (i.e. AaBb could produce AB, Ab, aB and ab gametes).

Question 3

A both parents are heterozygous.

A 3 : 1 ratio in offspring occurs when two heterozygous individuals are crossed. **B** is incorrect because the two different phenotypes in the offspring are the result of different genotypes. **C** is incorrect because both parents must be heterozygous, meaning three in every four offspring would have the dominant phenotype. **D** is incorrect because if the alleles display incomplete dominance, both parents must have the same genotype (because they have the same phenotype). If both parents are homozygous, all offspring will be the same; if both parents are heterozygous, a 1 : 2 : 1 ratio will be present in offspring.

Question 4

B It is expressed in heterozygous individuals.

The trait is autosomal dominant, so individuals who are heterozygous (i.e. have one affected allele) will display Marfan syndrome. **A** is incorrect because the trait is autosomal, not sex-linked. **C** is incorrect because natural selection leads to disadvantageous traits becoming less common over time, even if they are inherited in a dominant fashion. **D** is incorrect because the trait is dominant, therefore only one mutated allele is required for expression of Marfan syndrome.

Question 5

D Incomplete dominance

Incomplete dominance occurs when a dominant allele does not completely mask the effects of a recessive allele; in other words, both alleles are expressed, resulting in an intermediate phenotype. In this case, the normal allele is likely to be partially dominant, resulting in the production of receptors. An individual who is heterozygous (i.e. has only one of these dominant alleles) would produce roughly half the number of receptors compared with a homozygous dominant individual. **A** is incorrect because sex-linked inheritance occurs when the genes involved are located on sex chromosomes. **B** and **C** are incorrect because when there are two alleles, and one is completely dominant over the other, there are only two possible phenotypes; if this was the case, there would be either normal levels of receptor, or no receptor.

Question 6

C 4 genotypes and 3 phenotypes

A female who is heterozygous mallard must have the genotype Mm^d. A cross between this female and an M^Rm^d male will produce a genotypic ratio of 1 M^RM : 1 M^Rm^d : 1 Mm^d : 1 m^dm^d. M^RM and M^Rm^d will both display the restricted phenotype, Mm^d will be mallard and m^dm^d will be dusky. **A**, **B** and **D** are incorrect because there are four possible genotypes and three phenotypes.

Question 7

C 50%

The male parent's genotype is M^Rm^d, so the male's phenotype is restricted, because of the presence of an M^R allele. The phenotypic ratio is 2 restricted : 1 mallard : 1 dusky. **A**, **B** and **D** are incorrect because two of every four offspring will be restricted; that is, 50% will have the same phenotype as the male parent.

Question 8

C 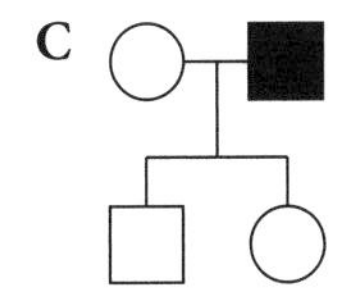

A male with a mitochondrial disease typically will not pass the disease on to any of his children. This is because paternal mitochondria are usually destroyed upon fertilisation, leaving only maternal mitochondria to be inherited by the offspring. The mother is unaffected, and hence the children would also be unaffected. **A** and **B** are incorrect because if their mother was affected, both children would inherit the affected mitochondria, and would themselves be affected. **D** is incorrect because a male with a mitochondrial disease will not pass the disease on to any of his children. The mother is unaffected; hence the children would also be unaffected.

Question 9

B Males have a grandfather and can have grandsons.

A male can produce a daughter when a haploid sperm combines with an egg. This daughter could then produce a son from her unfertilised egg. This would be the original male's grandson. By this same process, the male would have a grandfather. **A** is incorrect because sons can only be produced from an unfertilised egg. Females produce eggs. Thus, males can only produce daughters. **C** is incorrect because a female is formed when haploid sperm and egg combine; she would therefore be diploid and have twice the number of chromosomes that a male has, not half. **D** is incorrect because only diploid females would undergo meiosis to produce haploid gametes. Males are already haploid, so would undergo mitosis (not meiosis) to produce haploid sperm.

Question 10

D 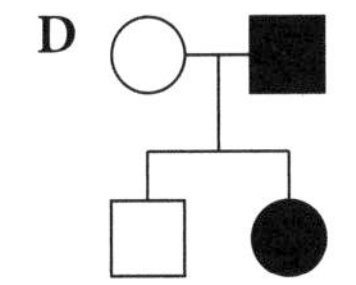

The affected male must pass his affected X chromosome onto his daughter. Because the condition is recessive, the affected daughter could have inherited a second affected X chromosome from her unaffected mother, who is a carrier. The mother passes her unaffected X chromosome onto her unaffected son. The son inherits a Y chromosome (which is unaffected) from his affected father. **A** and **B** are incorrect because if the disease is recessive, the affected daughter in both cases must have two affected X chromosomes, one of which must have been inherited from her father. If this were the case, her father, who only has one X chromosome, would also be affected. **C** is incorrect because the affected mother and father would each have to pass an affected allele onto their daughter, who would be affected as a result.

Short-answer solutions

Question 11

a 1 tabby:1 tan (1 mark)

b In codominance, heterozygous individuals express both alleles (1 mark) – tabbies have some orange hairs and some black hairs. In contrast, incomplete dominance results in heterozygotes having a 'blend' of the two phenotypes that is different from the two parent phenotypes (1 mark). For example, a snapdragon with one red allele and one white allele would have pink flowers (1 mark).

Question 12

a A place where the DNA differs between individuals by a single base pair. (1 mark)

b A condition that is caused by a gene carried on the non-sex chromosomes, the autosomes. (1 mark)

c **i**

Allele	Frequency
T	0.44
t	0.56

(2 marks)

ii Both parents must be heterozygous Tt (1 mark). The pups with normal tails are tt and must inherit a t from each parent (1 mark). The pup that dies at birth must have had a TT genotype, meaning it inherited a T from each parent (1 mark).

A Punnett square could be used as a justification, provided it is correctly completed.

Question 13

Female offspring must inherit one X chromosome from their father and one from their mother. In this cross, females will inherit either a normal or an affected allele from their mother, but they will always inherit a normal X chromosome from their father. This means that they cannot be homozygous for Menkes disease, and therefore (because it is a recessive disease) cannot inherit the disease. (2 marks)

Male offspring must inherit their Y chromosome from their father, and their X chromosome from their mother. Because their mother is heterozygous, male offspring have a 50% chance of inheriting the Menkes allele. Without a second X chromosome (carrying a normal allele), any males who inherit the affected X chromosome from their mother will have Menkes disease (2 marks). This is shown in the Punnett square below.

Key:
X^M = normal
X^m = Menkes disease

		Father	
		X^M	Y
Mother	X^M	X^MX^M	X^MY
	X^m	X^MX^m	X^mY

(Punnett square 1 mark)

2 marks are allocated to the explanation of the female offspring, and 2 marks for the explanation of the male offspring. 1 mark is allocated to the use of a Punnett square (with a key) to support the explanations.

Question 14

a (3 marks)

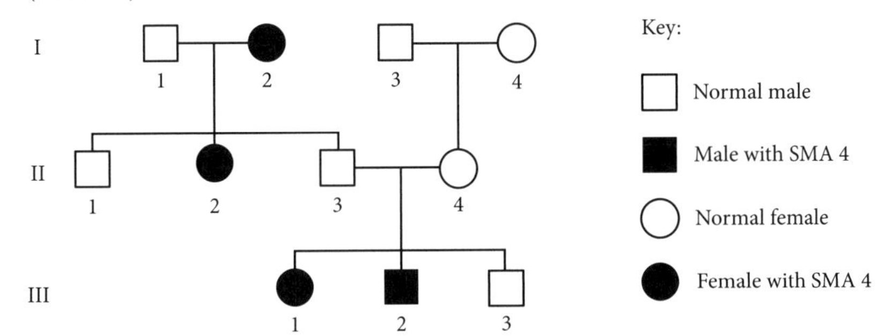

1 mark is allocated to the correct layout with the correct shading for affected and unaffected members, 1 mark for numbering individuals correctly, and 1 mark for a correct key.

b SMA 4 must be autosomal recessive (1 mark). For III1 and III2 to be affected despite having two unaffected parents, it must be recessive, with II3 and II4 both being carriers (1 mark). If it were sex-linked, males II1 and II3 would inherit their one X chromosome from their affected (homozygous recessive) mother and would therefore have SMA 4. Because they don't have the disease, it must be autosomal (1 mark).

1 mark is for correctly identifying that SMA 4 is autosomal recessive; 2 marks are for the justification. You must refer to specific individuals by number in order to score full marks. General comments like 'it skips a generation' or 'more people are unaffected than affected' are not adequate justifications for 2 marks.

Question 15 ©NESA 2019 MARKING GUIDELINES SII Q28

a

Alleles from father
H, r (1 mark)

Alleles from mother
h, R (1 mark)

The stimulus states that the patient's mother does not have either disease and is homozygous for both genes. This means the patient must have inherited normal 'h' and 'R' alleles from his mother. This also means that he inherited both 'defective' alleles from his father. For Huntington's, the defective allele is the dominant 'H' allele, while for Stargardt disease the defective allele is the recessive 'r' allele.

b (4 marks)

Homologous pair of chromosome 4 before crossing over	H H / r r		h h / R R	
Homologous chromosomes after crossing over and separation	H H / r R		h h / r R	
Gametes	H / r	H / R	h / r	h / R

KEY
H – Huntington's
h – Non-Huntington's
R – Healthy retina
r – Stargardt disease retina

The completed diagram must include a key (1 mark). It must model the initial genotype of the homologous pair from part **a** as well as the correct sister chromatids before crossing over (1 mark), the process of crossing over (1 mark) and the alleles present on the four gametes (1 mark).

Test 5: Inheritance patterns in a population

Multiple-choice solutions

Question 1

C The complete set of genetic information within an organism

An organism's genome refers to all the genetic information of an organism, including the coding DNA within genes as well as the non-coding DNA within and between genes. **A** is incorrect because it refers to proteins, rather than the genetic information itself. **B** is incorrect because it excludes the non-coding DNA between genes. **D** is incorrect because it excludes genes that are not 'switched on' in any given cell.

Question 2

C It can be used to determine the risk of inheriting a genetic disease.

DNA sequencing can determine whether a gene contains mutations that may be linked to a genetic disease. **A** and **B** are incorrect because DNA sequencing can be carried out on both prokaryotic DNA and non-coding DNA. **D** is incorrect because DNA sequencing does not amplify DNA; PCR is the technology used for this purpose.

Question 3

A DNA sequencing

DNA sequencing can be used to measure evolutionary relatedness because the more similar the base sequence of DNA of two species, the more closely related they are. **B** is incorrect because different species may have similar mating behaviours. **C** is incorrect because amino acid sequences may indeed vary from one species to the next; however, because of the degeneracy of the DNA code, the amino acid sequences of proteins may be the same even though the DNA that codes for those proteins may be different. Therefore, DNA sequencing is more useful than amino acid sequencing. **D** is incorrect because morphological features can vary greatly, even between two members of the same species.

Question 4

B study the evolutionary history of humans.

DNA fragments can be amplified and used to trace evolutionary history. This is possible because DNA mutates at predictable rates. The more similar the DNA of two species, the less time has passed since the two species diverged. **A** is incorrect because an organism cannot be cloned from only fragments of DNA. **C** is incorrect because, even if it were possible, it would be unethical to conduct such an experiment. **D** is incorrect because, although modern-day humans and chimpanzees have a common ancestor, one did not evolve from the other.

Question 5

D None of the males can be excluded.

The child, being blood type O, must have the genotype i^oi^o, and must have inherited an i^o allele from each parent. **A** is incorrect because male 1 would have the genotype i^oi^o, meaning he would definitely pass on an i^o allele to the child. **B** is incorrect because male 2 could have the genotype I^Ai^o, meaning it would be possible for him to pass on an i^o allele to the child. **C** is incorrect because male 3 could have the genotype I^Bi^o, meaning it would be possible for him to pass on an i^o allele to the child. Therefore, no male can be eliminated.

Question 6

C Male 3

Each band in the child's fingerprint should also be visible in the fingerprint of at least one of their parents. The band at 5500 is also seen in the fingerprint of the mother, but the band at 1500 is not. Thus, this band must be visible in the fingerprint of the father. **A** and **B** are incorrect because neither of these males has a band at 1500. **D** is incorrect because male 3 has a band at 1500; therefore, it is possible to tell that male 3 is the father.

Question 7

B An increase in the age at which women conceive

The risk of Down syndrome increases with maternal age. Therefore, an increase in the age at which women conceive would lead to more babies being born with Down syndrome. **A** is incorrect because an increase in genetic screening would either have no impact, or would decrease the incidence. This is because greater screening rates would mean a higher rate of detection, and some of these pregnancies would likely be terminated before birth. **C** is incorrect because the life expectancy does not impact on incidence, which is the number of new cases diagnosed. It would impact on prevalence, or the number of people with Down syndrome at a given time. **D** is incorrect because an increase in termination rate would decrease incidence, not increase it.

Question 8

A It might cause changes in gene expression.

Non-coding regions of DNA do not provide the instructions for the production of polypeptides. Although the function of many non-coding regions is not completely understood, it is clear that at least some of the non-coding DNA plays a role in regulating the expression of genes. **B** and **D** are incorrect, because non-coding DNA does not provide instructions for making polypeptides and proteins from amino acids. **C** is incorrect because the nucleotide sequence of non-coding DNA is not reflected in the mature mRNA.

Question 9

A

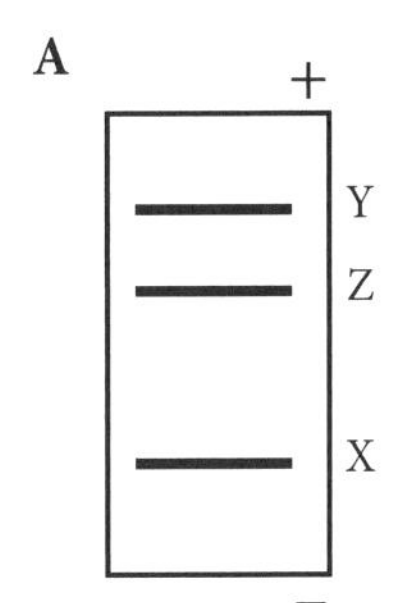

DNA fragments are added at the negative electrode and migrate towards the positive electrode. The speed of migration is proportional to the length of the fragment, with smaller fragments migrating more quickly that larger ones. **B** and **D** are incorrect because fragment Y, being the smallest, would move the fastest, and would therefore be furthest from the negative terminal. **C** is incorrect because the size difference between X and Y is not equal to that of Z and Y. Therefore, the fragments would not be equally spaced as in **C**.

Question 10

D Human chromosome 2 is the result of the fusion of two chromosomes in the common ancestor of all primates.

The banding patterns of the chromosomes match closely, despite the non-human primates having two separate chromosomes and the human only having one. This suggests that ancestors of these three species either had two chromosomes that fused together in humans, or one chromosome that split into two in non-human primates. **A** is incorrect because humans did not evolve from chimpanzees; they have a common ancestor. **B** is incorrect because the banding pattern of humans is more similar to that of chimpanzees than that of gorillas. This indicates that humans are more closely related to chimpanzees than to gorillas. **C** is incorrect because non-human primates did not evolve from humans; they have a common ancestor.

Short-answer solutions

Question 11

a Suspect 2 is the likely perpetrator (1 mark), as their DNA profile matches the sample found at the crime scene (1 mark).

b The victim, as well as the perpetrator, may have left DNA at the crime scene (1 mark). A sample from the victim is needed to exclude the victim as the source of the evidence in question (1 mark).

Question 12

a Nuclear DNA is inherited from both mother and father, whereas mtDNA is usually inherited from the mother only (1 mark).

b Inheritance of mtDNA is easier to trace from one generation to the next. There is no 'crossing over' as in meiosis and no 'mixing' of maternal and paternal DNA during fertilisation, as there is with nuclear DNA (1 mark). This makes it easier to trace inheritance patterns and determine evolutionary relationships between the ancestors of modern humans (1 mark).

Because the mtDNA is passed on in its entirety, any changes from one generation to the next can be directly attributed to mutations (1 mark). Mutations in DNA occur at a predictable rate, so changes in mitochondrial DNA can be used to estimate the time since two closely related *Homo* species diverged (1 mark).

To achieve full marks, two reasons are required. For each reason, 1 mark is allocated to stating the feature of mtDNA, and 1 mark to explaining why this feature makes it more useful than DNA for studying human evolution.

Question 13

Child 3 will have sickle cell anaemia (1 mark) because the mother, father and Child 2 are all heterozygous/unaffected (Aa) and this is depicted by the two bands on the DNA profile (1 mark). Child 1 is also unaffected and has only one band, and therefore must be homozygous normal (AA). Child 3 displays one band in a different location from Child 1, but the band is common to the heterozygous family members. Therefore, Child 3 must be (aa) recessive and will have sickle cell anaemia (1 mark).

1 mark is allocated to stating that Child 3 will have sickle cell anaemia. 2 marks are allocated to a full justification. A full justification should refer to all members of the family, and how the bands of their profile link to their genotype and provide evidence for the genotype of Child 3.

Question 14

a (4 marks)

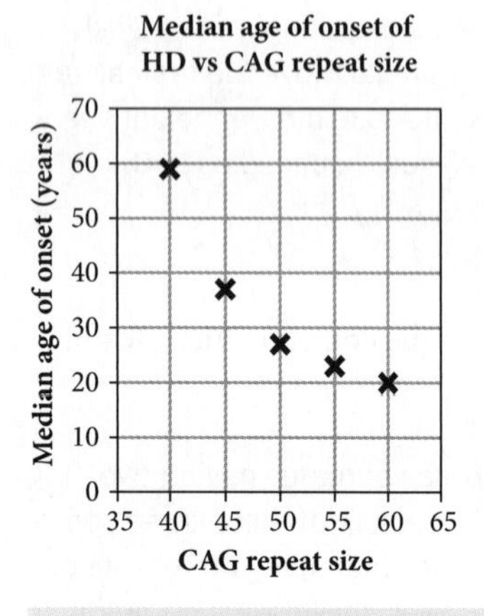

The independent variable is the CAG repeat size, so it should be on the *x*-axis, and the median age in years should be on the *y*-axis (1 mark). Each axis should have a linear scale (1 mark). The data should be plotted accurately (1 mark) and the graph should have a suitable title (1 mark).

b As the number of CAG repeats increases, the age of onset decreases (1 mark). There is a sharp decrease in age of onset from 40 to 50 repeats. It begins to flatten out as the CAG repeat size increases above 55 (1 mark).

Note that the question says describe the trend, not just identify, and this should be reflected in your answer. The trend is clearly non-linear.

c For a study to produce reliable data, it must use a large sample size. Because Huntington's disease is rare (1 mark), collaboration must occur on a large scale to gather a large enough sample size to ensure any conclusions drawn are reliable (1 mark).

Question 15

a The more frequent the heterozygous genotype, the higher the incidence. (1 mark)

b **i** Fewer bacteria enter cells that are heterozygous for the CF allele than enter normal cells. (1 mark)

ii The graph shows that having a heterozygous genotype provides a level of protection against pathogenic bacteria; for example, for species A, twice as many bacteria enter normal cells than heterozygous cells (1 mark). This 'heterozygous advantage' means that, despite the homozygous recessive genotype being detrimental, the CF allele persists in the population. As a result of natural selection, in areas where these bacterial infections were once common, individuals with the heterozygous genotype would be more likely to survive (1 mark), passing their CF alleles onto their offspring (1 mark). In some cases, this would result in an individual with CF, while in other cases, this would result in offspring who are heterozygous (1 mark). Thus, the CF allele, and the heterozygous genotype, increase in frequency.

You need to address the role of natural selection in this question using information from the graph (i.e. the fact that being heterozygous provides an advantage regarding bacteria entering the cells of the gut). The term 'heterozygous advantage' is not necessary to achieve full marks.

c Large-scale collaboration allows laboratories and researchers to pool their resources (1 mark). This makes research cheaper, prevents unnecessary overlap of research and increases the efficiency of the research process (1 mark), allowing important discoveries to be made sooner.

Collaboration across multiple countries (in the case of CF, 30 countries) allows researchers to gather a more comprehensive set of data (1 mark). This allows for trends across different ethnic groups (such as the CF incidence in the table) to be revealed; without collaboration, these patterns may not have been evident (1 mark).

To score full marks, two appropriate reasons must be provided, with a valid explanation for each. Reasons other than those stated here are acceptable, provided they are logical.

Test 6: Mutation

Multiple-choice solutions

Question 1

B Aneuploidy

Aneuploidy occurs when an extra or missing chromosome leads to a chromosome number that is not an exact multiple of the haploid number. In the case of Edwards syndrome, the chromosome number is $2n + 1$. **A** is incorrect because polyploidy refers to entire extra sets of chromosomes (e.g. $3n$ or $4n$). **C** is incorrect because a point mutation refers to a mutation affecting a single gene rather than an entire chromosome. **D** is incorrect because insertion mutations occur when one or more nucleotides are added to a DNA sequence, or when a section of a chromosome is translocated. Neither of these insertions results in a change in chromosome number.

Question 2

A silent mutation.

A silent mutation occurs when a base substitution in DNA results in an mRNA codon that still codes for the same amino acid. **B** is incorrect because a missense mutation occurs when a substitution results in a different amino acid. **C** is incorrect because a nonsense mutation occurs when a base substitution results in a stop codon, leading to a shortened and probably non-functional protein. **D** is incorrect because a frameshift mutation occurs when one or more bases is inserted or deleted, resulting in a change in the reading frame of the ribosome and a change in many amino acids.

Question 3

A It will have an effect on many genes.

Deletion of large sections of a chromosome will affect the many genes that are located within the deleted region. These genes will be under-expressed in the resulting cell. **B** is incorrect because many codons will be missing. **C** and **D** are incorrect because errors during transcription and translation would affect the mRNA and subsequent polypeptide, not the chromosome itself.

Question 4

C genetic drift.

Genetic drift occurs when a chance event randomly changes the frequency of the alleles in a population. The bottleneck effect is an extreme example of genetic drift. **A** is incorrect because mutation introduces new alleles into the population, and a change in frequency of these alleles generally occurs over a long period of time. **B** is incorrect because gene flow describes the transfer of genetic material from one population to another, which is not occurring in this instance. **D** is incorrect because natural selection is not random; instead, changes in the allele frequency depend on which alleles are advantageous.

Question 5

C They are found both between and within genes.

Non-coding DNA is found between genes (intergenic regions) and also within genes (introns). **A** is incorrect because mutations often occur in non-coding regions of the DNA. **B** is incorrect because only coding DNA (the exons within genes) is expressed as polypeptides. **D** is incorrect because the vast majority of DNA is non-coding.

Question 6

D the reshuffling of alleles as a result of sexual reproduction.

Meiosis and fertilisation lead to alleles being shuffled. This causes variation in the genotypes of individuals, even if they are closely related. **A** is incorrect because the environmental factors that influence phenotype do not usually impact on the genetic code itself. **B** is incorrect because it is not each new mutation, but the accumulation of these mutations over many generations, that has led to great variation in humans. **C** is incorrect because the human population is large.

Question 7

B Translocation

Translocation mutations occur when a piece of one chromosome breaks off and attaches to another chromosome. **A** is incorrect because aneuploidy occurs when there are extra or missing chromosomes; here, there is still the correct chromosome number. **C** and **D** are incorrect because point mutations and frameshift insertions only affect a single gene, not a large section of a chromosome as shown here.

Question 8

D Individual A: male. Individual B: female

It is the presence of the *SRY* gene, not the Y chromosome itself, that leads to male development. Individual A will be male, because of the presence of the *SRY* gene, and individual B will be female because of the absence of the *SRY* gene. Therefore **A, B** and **C** are incorrect.

Question 9

C It causes base substitutions during DNA replication.

5-BrU causes base substitutions, resulting in a T–A pair being replaced by a 5-BrU–G pair. **A** is incorrect because thymine or cytosine dimers cause these kinks; 5-BrU is structurally similar to thymine, so it is unlikely to change the overall shape of DNA. **B** is incorrect because chromosomal rearrangements affect large sections of a chromosome, whereas 5-BrU affects individual base pairs. **D** is incorrect because changes in the reading frame only occur when bases are inserted or deleted from the DNA sequence, not when one is substituted for another.

Question 10

B The SNPs are part of DNA that is not expressed.

It is likely that the SNPs shown are located in non-coding DNA – that is, in DNA that is not expressed. This is because about 99% of DNA does not code for polypeptides. **A** is incorrect because a change in one nucleotide can result in a change in an amino acid, which can lead to a change in phenotype (e.g. sickle cell anaemia). **C** is incorrect because these codons code for different amino acids. **D** is incorrect because SNPs are a change in a single base pair, not a single nucleotide.

Short-answer solutions

Question 11

a A mutagen is an environmental agent that is able to alter DNA and cause mutations. (1 mark)

b Ultraviolet radiation (1 mark) from sunlight causes adjacent thymine bases on the same strand of DNA to attach to each other, rather than to their complementary base pair on the opposing strand. This causes a structural kink in the DNA that prevents DNA replication and transcription (1 mark).

Question 12

a Only one X chromosome is present, where there should be two. (1 mark)

b During meiosis, the sex chromosomes may fail to segregate (1 mark). This is called non-disjunction and would result in one gamete containing two sex chromosomes, and another gamete containing no sex chromosomes (1 mark). When a gamete containing no sex chromosomes combines with a normal gamete containing an X chromosome (1 mark), the result would be an offspring with Turner syndrome.

To score full marks, the term 'non-disjunction' is not required, provided the process is outlined to explain how a gamete without a sex chromosome could be produced. Your answer must also include the role that fertilisation plays in this gamete becoming an offspring who has Turner syndrome.

Question 13 ©NESA 2021 MARKING GUIDELINES SII Q24

a It is a somatic mutation (1 mark) because it is a dominant trait but not present in the parents (1 mark) or the offspring of the affected male. If it were a germ-line mutation, which occurs in the cells that form gametes, it is most likely that it would be seen in the offspring (1 mark).

b Mutation B occurred after twin formation, therefore affecting only Twin 1. Because it occurred prior to formation of the germ cells it will be evident in both the germ cells and the somatic cells of Twin 1 (1 mark), and through the germ cells may be passed to the offspring (1 mark). Because Mutation C occurs in the cells leading to somatic cells in Twin 2 only, it will be evident in the somatic cells of Twin 2 (1 mark) but not in the germ cells and can't be passed on to offspring (1 mark).

To score full marks you must address both twins and their potential offspring. You must also show that you understand the impact of the timing of the mutation on its effect – that is, whether the mutation occurs before or after cells differentiate into germ or somatic cells will determine whether the mutation affects the twins' offspring or just the twins themselves.

Question 14

a Frameshift (1 mark)

Insertion is also correct.

b (4 marks)

Normal HEXA DNA	GCA	TAT	AGG	ATA	CGG	GGA	CTG	...
Normal HEXA mRNA	CGU	AUA	UCC	UAU	GCC	CCU	GAC	...
Normal HEXA polypeptide	Arg	Ile	Ser	Tyr	Ala	Pro	Asp	

Tay–Sachs HEXA DNA	GCA	TAT	AGA	**TAG**	GAT	ACG	GGG	ACT	G...
Tay–Sachs HEXA mRNA	CGU	AUA	UCU	**AUC**	CUA	UGC	CCC	UGA	C...
Tay–Sachs HEXA polypeptide	Arg	Ile	Ser	**Ile**	Leu	Cys	Pro	Stop	

Note that the question states that the DNA code provided is the 'template strand'. This means the mRNA should be the complement to the DNA. This is awarded 1 mark. If the base pairing between DNA and mRNA is correct (i.e. A–U, C–G), a second mark is awarded. A third mark is for the correct amino acid sequence of the normal HEXA polypeptide, and the final mark is for the correct amino acid sequence in the Tay–Sachs HEXA polypeptide.

c The addition of the four bases causes a change in reading frame of the ribosome. As shown in the table, a premature stop codon is reached as a result of the mutation. This stop codon would result in a polypeptide that is shorter than normal (1 mark).

This would alter the three-dimensional structure of the beta-hexosaminidase enzyme (1 mark). The change in the protein shape would make it unable to break down GM2 ganglioside (1 mark), which would build up in the brain and spinal cord cells and lead to the symptoms stated in the question.

Question 15

a The total collection of genes in a population at any one time; consists of all the alleles in all the individuals in the population. (1 mark)

b Because of the small size of each population, the gene pool of each population and the numbat species as a whole would be small. Mutations, or changes to the genetic code (1 mark), will introduce new alleles and increase the size of the gene pool of the numbat populations (1 mark). If a mutation has a positive effect on survival, it has the potential to have a large impact on the gene pool through the process of natural selection.

Gene flow is the transfer of genetic material from one population of numbats to another, usually through migration (1 mark). This would increase the size of each population's gene pool. Gene flow would have minimal impact on the gene pool naturally, because of isolation of the numbat populations. Human intervention, such as that seen in the translocation of the males and females in 2005 and 2010 from Dryandra to Boyagin populations, would increase the size of the Boyagin gene pool, especially if these individuals went on to breed with others (1 mark).

Genetic drift is the random change in the frequency of an existing gene variant in a population, by chance events (1 mark). Fire and timber harvesting events would result in genetic drift. The frequency of certain alleles in the numbat population will increase and decrease over time because of these events. The effects of genetic drift would be especially profound in the numbat populations because of the small size of these populations (1 mark).

Mutation and gene flow have the potential to lead to small but important increases in the gene pool of the numbat populations. Genetic drift may have a large impact on allele frequency changes because of the small population size (1 mark).

2 marks are allocated to each process (mutation, gene flow and genetic drift): 1 mark is awarded for a description of the process, and one for explaining how this process affects (or could affect) the numbat populations. In order to score full marks you must evaluate or make a judgement about these effects; ideally this evaluation is integrated throughout the explanation of each process (as shown in the sample above), rather than a single statement addressing the overall impact of the three processes.

Test 7: Biotechnology

Multiple-choice solutions

Question 1

D The removal of weeds by humans to reduce their impact on native species

Biotechnology is an application that uses living things, including plants, animals or microbes to benefit humans. The removal of weeds does not require the use of another species, and it benefits native species rather than humans directly. **A** is incorrect because the use of bacteria to prevent disease in humans is an example of biotechnology. **B** is incorrect because the use of bacteria to produce food for humans is an example of biotechnology. **C** is incorrect because the use of fungi to make drugs used to treat disease in humans is an example of biotechnology.

Question 2

D The variety of organisms within a habitat or region

Species diversity relates to the number of species present. **A** is incorrect because this describes the genetic diversity within a species. **B** is incorrect because this describes ecosystem diversity. **C** is incorrect because this describes the genetic make-up of an individual.

Question 3

A Cloning

Vegetative propagation produces offspring that are genetically identical to the (single) parent plant; that is, the offspring are clones of the parent and of each other. **B** is incorrect because hybridisation involves crossing two genetically different parents. **C** is incorrect because selective breeding also requires two parents. **D** is incorrect because artificial pollination involves humans transferring pollen from one flower to the stigma of another, which is not occurring here.

Question 4

C Lack of biodiversity

Because most potatoes were the same variety, produced by cloning, there was little diversity in the potato crops. They were not resistant to the fungus, so most of the crops failed. **A** is incorrect because mutation would have increased the variety within the potatoes; new mutations may have provided a level of resistance to the fungus. **B** is incorrect because genetic drift eliminates individuals from the population at random, not because they are susceptible to a certain disease. **D** is incorrect because the pathogen is a fungus, not an insect pest, so insecticides would not have been used.

Question 5

B Gene flow from the crops to the wild plants

Gene flow is the transfer of genetic material from one population to another, in this case, from the transgenic crop to the wild relatives. **A** is incorrect because crossing over in the wild plants cannot account for the presence of the transgene in wild plants. **C** is incorrect because genetic drift occurs when a chance event randomly changes the frequency of the alleles within a single population, not across two populations as seen here. **D** is incorrect because a mutation cannot introduce a gene from another species, it can only alter genes that are already present.

Question 6

C It could relieve pressure on wild fish stocks as a source of omega-3 oils.

GM canola will provide a new source of omega-3 oils. Fish are currently the primary source, so commercial use of the GM canola will mean fewer fish will need to be caught to meet demand. **A** is incorrect because omega-3 oil does not cure type 2 diabetes, it merely reduces the risk of developing it. **B** is incorrect because if it is grown commercially, more land would be required for canola farming. **D** is incorrect because if land is cleared to farm GM canola, this could result in a reduction in biodiversity.

Question 7

D Ethical considerations

Ethical considerations are concerned with morality and the balance between benefit and harm. **A** is incorrect because a discussion about economic impacts would address the financial costs and potential profits, not the welfare of the horses involved. **B** is incorrect because social implications would deal with the broader potential impacts of the treatment on society, rather than on the individual receiving the treatment. **C** is incorrect because a discussion about cultural influences would address how an individual's customs, religion etc. may influence their opinion or use of the treatment.

Question 8

D maintain strains with characteristics that may prove useful for future commercial varieties.

Wild strains should be preserved because they may have characteristics that could be useful in future, such as resistance to future pathogens, or being better suited to a future climate. **A** is incorrect because many places already have ideal rice-growing conditions. **B** is incorrect because the aim is to maintain genetic diversity, not reduce it. **C** is incorrect because farmers do not require approval to selectively breed wild strains; wild strains are not 'owned' by anyone.

Question 9

D It will reduce the number of people who die while waiting for an organ transplant.

There is currently a shortage of human donor organs; this technology will improve availability of organs for transplant. **A** is incorrect because rearing genetically engineered pigs in a highly regulated environment would potentially be more expensive than using a human donor. **B** is incorrect because, even if pigs are screened for known pathogens, there remains a risk that unknown pathogens may infect the recipient. **C** is incorrect because individuals may have ethical concerns about the use of pigs as a source of donated organs.

Question 10

D The transfer of genes between two species increases biodiversity.

A transgenic species has had DNA from another species inserted into its genome. This introduces new alleles into the population, increasing the biodiversity of that species at a genetic level. **A** is incorrect because no new genes are created, they are just transferred from one species to another. **B** is incorrect because no genes are removed from any species. **C** is incorrect because an organism that has received genes from another organism of the same species is not transgenic.

Short-answer solutions

Question 11

a Biotechnology is an application that uses living things, including plants, animals or microbes, to make or modify products or processes for the benefit of humans. (1 mark)

b For example: For thousands of years, humans have used the process of microbial fermentation to preserve milk and produce yoghurt and cheese (1 mark). It is considered a biotechnology because animals (e.g. cattle, goats) were used to obtain the milk, and bacteria were used to turn it into other products (1 mark). This process is of great benefit to humans, because it allows for the preservation and production of food (1 mark), hence it is considered to be a biotechnology.

Question 12

Artificial pollination was used to carry out selective breeding of papayas. Pollen from a male plant with favourable characteristics (such as pleasant flavour and appearance) would have been used to pollinate the flowers of female plants who also had these traits (1 mark). This would have decreased the genetic diversity of the population, because only those with favourable characteristics were allowed to reproduce in large numbers (1 mark).

Genetic modification of the papaya would have initially increased the genetic diversity of the papayas by introducing new genes to the gene pool (1 mark). However, because the GM papayas account for approximately 90% of Hawaii's papayas, it is clear that they have been used on a large scale, reducing the genetic diversity of the overall population (1 mark).

For each genetic technique, 1 mark is allocated to an outline of the genetic technique, and 1 mark is allocated to explaining how it affects the genetic diversity of the papayas.

Question 13

a The farmer used selective breeding to increase the life span of the animals. (1 mark)

b Based on the data, this is a very effective use of biotechnology (1 mark), because the number of offspring increased but the percentage of early deaths decreased (1 mark). Of 53 offspring in generation 1, 27 died early, whereas only 10 of 62 offspring died early in generation 3 (1 mark). It appears that brown offspring have better survival rates, as shown by the proportion of males of each colour surviving to breeding age. These beneficial characteristics are selected for and appear to be successfully passed on to the next generation (1 mark).

Note that the question specifically requires an assessment. This means a judgement must be made about the effectiveness of the biotechnology. The question also says, 'using the farmer's data'. In order to score full marks, you must use more than one trend from the data.

Question 14

	GM sorghum	Bt eggplant
Potential benefit to society	Higher protein content and increased digestibility mean more nutritious food for animals who consume GM sorghum. Because the nutrient content is higher, less animal feed will be needed, reducing the strain on current food supplies. (1 mark)	Bt eggplant requires 92% less pesticide use. This benefits society because farmers, factory workers and consumers will be less likely to suffer from the detrimental effects of pesticide exposure. (1 mark)
Social implications	Farmers in developing countries may not have the financial means to keep buying GM canola seeds, so there may be unequal access to this technology; those who need it most may be unable to benefit from it. (1 mark)	The increased yield and reduced reliance on pesticides makes farming eggplant in India more profitable, increases the livelihoods of farmers and has a positive impact on their wellbeing. (1 mark)
Future direction for research	How to prevent gene flow between transgenic sorghum lines and wild relatives, either through genetic engineering or through crop management practices. (1 mark)	Which crop management practices best prevent insect developing resistance to the BT toxin. (1 mark)

To achieve 1 mark in each component, you must elaborate on the information provided; it is not sufficient to merely restate information from the stimulus in each box. Note that social implications relate to the effects of biotechnologies on the social fabric of the community and the wellbeing of individuals and families. Social implications involve a person's emotions, opinions or behaviours being affected, either intentionally or unintentionally. There may be some overlap between benefits to society and social implications.

Question 15

Biotechnology is an application that uses living things, including plants, animals or microbes, to produce products or carry out processes that benefit humans. Such biotechnologies often lead to ethical issues because they involve the competing interests of many parties (1 mark). In a utilitarian approach, all the benefits, as well as any potential harm, that may result from the development of a new biotechnology must be considered.

The use of transgenic mice as models for Alzheimer's disease (AD) has the potential to benefit millions of people. Experiments with transgenic mice may lead to the discovery of novel treatments that can slow or even reverse the progression of AD (1 mark). This could improve the quality of life of millions of people, reducing the cost to the healthcare system and keeping these individuals working and living independently (1 mark).

These benefits must be weighed against the potential harm that the technology may inflict. The successful creation of a single transgenic mouse may have been the culmination of hundreds of unsuccessful attempts, resulting in the death of thousands or tens of thousands of mice. 'Successful' attempts to induce AD in test mice would result in the mice suffering from the devastating effects of AD. Mice are sentient beings, capable of feeling emotions such as pain and fear. One must consider whether it is ethical to intentionally inflict so much pain and suffering on other organisms in this way, even though it is for a good cause (1 mark).

Genetically modified crops, such as golden rice and BT corn, have the potential to eliminate world hunger and nutritional disease (1 mark). They have the potential to eliminate or at least reduce our reliance on pesticides. This in turn reduces harm to humans and animals who may otherwise be exposed to toxic or carcinogenic chemicals (1 mark).

However, there are risks associated with the development of new GM crops. The potential for 'contamination' of surrounding native species could lead to potential disruption of delicate ecosystems, leading to a reduction in biodiversity if species or strains are lost (1 mark). The legalities of GM crop ownership have the potential to increase inequalities between high-income and low-income countries.

When developing new biotechnologies, a utilitarian approach requires one to consider both the potential benefits and the possible harms for all parties involved. This includes the organisms used in the biotechnology, the humans who benefit from their use, and any other humans or organisms that may be affected a result of their development and future use.

Note that there are many ways to approach this question that could score full marks. The example above is one such example. The question says 'biotechnologies', plural. This means that in order to score full marks, you must explain **two** biotechnologies; one must be the AD example provided. For each biotechnology, you must discuss (i.e. points for/benefits and points against/harms). You must then relate these to the theory of utilitarian ethics (i.e. the balance between benefits and harms).

Test 8: Genetic technologies

Multiple-choice solutions

Question 1

B Artificial pollination

Artificial pollination involves humans transferring pollen from one flower to the stigma of another. **A** is incorrect because the process the scientist carried out does not result in identical copies of a gene being produced, as in gene cloning. **C** is incorrect because artificial insemination is only possible in animals. **D** is incorrect because whole-organism cloning in plants does not involve gametes.

Question 2

C Parent plants with specific traits, such as larger buds, were bred together over many generations.

Selective breeding can be used to produce varieties with different characteristics. **A** is incorrect because this is a form of artificial selection, not natural selection, because it has been carried out by humans. **B** is incorrect because the question states that the process has been carried out over the past 2000 years, and transgenics is a relatively recent technology. **D** is incorrect because, although different environments may lead to some phenotypic variation, it is unlikely that evolution could have produced such vastly different varieties over a relatively short time period.

Question 3

C Tissue culture

The diagram shows the technique of plant cloning using tissue culture. **A** is incorrect because there is no transfer of genes from another species. **B** is incorrect because a whole organism is being cloned, not just a gene. **D** is incorrect because no pollination has taken place; tissue culture uses somatic cells, rather than the transfer of gametes, as in artificial pollination.

Question 4

A It is a form of selective breeding.

Hybridisation within a species involves humans deliberately selecting specific parent individuals and crossing them to produce offspring; thus, it is considered a form of selective breeding. **B** is incorrect because the two parents are members of the same species, so their offspring will be viable (able to produce offspring of their own). **C** is incorrect, because it will lead to gene flow between the two populations, increasing the genetic diversity. **D** is incorrect because both parents and the offspring will all have the same chromosome number because they are members of the same species.

Question 5

D

Artificial insemination	Cloning
Fewer males are used to reproduce	All individuals have the same genotype

In artificial insemination, semen from a single male is often used to inseminate multiple females. In cloning, all offspring produced are genetically identical to each other, and to the parent. For these reasons, both processes can decrease genetic diversity. **A** is incorrect because in artificial insemination, fertilisation is not random. **B** is incorrect because cloning does not involve gametes. **C** is incorrect because male gametes used in artificial insemination are genetically different from one another.

Question 6

A It improves the chance of successful fertilisation.

Hand pollination improves the chance of successful fertilisation because the ideal amount of pollen can be applied directly to the stigma. **B** is incorrect because hand pollination is more time consuming than machine pollination. **C** is incorrect because the offspring produced by sexual reproduction (of which artificial pollination is an example) are genetically unique. **D** is incorrect because it does not involve the addition of genes from other species.

Question 7

C Instructions for the production of the bacterial enzyme phytase are incorporated into the genetic material of pigs.

Transgenics involves the introduction of genetic material from one species into the genome of an organism from another species, through artificial methods. **A** is incorrect because both breeds are from the same species and no foreign genes are introduced. **B** and **D** are incorrect because, although there is genetic manipulation involved, addition of genetic material from another species does not occur.

Question 8

B It generates new alleles that may be useful.

Mutation induction increases the mutation rate. Mutations are a source of new alleles, some of which could lead to phenotypic changes that are useful to humans. **A** is incorrect because mutation induction would speed up the process of evolution, by generating new variations that can be acted upon by natural selection. **C** is incorrect because these new variations increase the genetic diversity within the population. **D** is incorrect because mutation induction does not prevent naturally occurring, spontaneous mutations.

Question 9

D tusk length.

Domesticated pigs have shorter tusks than wild boars. This means that humans must have selected *against* domesticated pigs having long tusks. **A** and **B** are incorrect because domesticated pigs have larger mass and height than wild boars. This means that humans must have selected *for* large mass and height. **C** is incorrect because the litter size of domesticated pigs and wild boar is the same, so it is likely that litter size has not been the focus of selective breeding of pigs.

Question 10

D There is no equivalent of a uterine wall in which to implant the developing embryo.

Once formed, the embryo produced by SCNT must be implanted into the uterine wall of the surrogate mother so that it can continue to develop. **A** is incorrect because birds do possess somatic cells. **B** is incorrect because, although it would be difficult to enucleate the egg cell of a bird because of the size and opacity of the yolk, removal of the nucleus from a somatic cell of the bird to be cloned would present no such problem. **C** is incorrect because the developing embryo is nourished by the yolk of the egg.

Short-answer solutions

Question 11

	Artificial insemination	Artificial pollination
Processes	Sperm is collected from a selected male and artificially introduced into the reproductive tract of a selected female. Semen may be diluted and frozen and used to inseminate many females. (1 mark)	Pollen is collected from the stamens of one plant and artificially transferred (by hand or machine) to the stigma of either the same plant (self-pollination) or another plant (cross-pollination). (1 mark)
Outcome, including a specific example	Animals with desirable characteristics, e.g. cows with greater milk production, or sheep with finer wool. (1 mark)	Plants with desirable characteristics, e.g. wheat with drought tolerance and pest resistance. (1 mark)

There are many possible responses to this question.

Question 12

a PCR is used to amplify small amounts of DNA (1 mark), e.g. to turn a small amount of DNA found at a crime scene into sufficient amounts to be analysed by DNA fingerprinting (1 mark).

b Step 1: DNA is heated to separate DNA strands (1 mark).

Step 2: DNA is cooled so primers can bind to each end of the DNA (1 mark).

Step 3: DNA polymerase uses the template to produce a complementary strand using free nucleotides (1 mark).

c PCR is like a chain reaction because the process above is repeated many times, leading to an exponential increase in the number of DNA copies (1 mark). One strand is turned into two, two strands into four, four into eight and so on, so there is a doubling of the number of DNA copies with each round (1 mark). Each step leads to the next, as in a chain reaction.

Question 13

Somatic cell nuclear transfer (SCNT) would involve taking the nucleus from a somatic cell of the prize-winning cow and inserting it into an enucleated egg. The resulting embryo would then be implanted into a surrogate mother (1 mark). This technique is inefficient: it has a low success rate, and often many attempts are needed to produce a single clone (1 mark). However, the farmer has probably chosen this technique because their main aim is to preserve the genetics of the cow. SCNT guarantees that the offspring will be genetically identical to the prize-winning cow, and, as result, it is likely to have the same beneficial traits as the original cow (1 mark).

Embryo splitting would involve fertilising an egg from the prize-winning cow with sperm from a selected bull *in vitro*. The developing ball of cells would be divided to produce multiple embryos, which could then be implanted into surrogates (1 mark). An advantage of this technique is that multiple identical offspring are produced (1 mark). The drawback of this technique is that it is a form of sexual reproduction. This means that both the male and the female will contribute to the genetic make-up of the offspring. There is no guarantee that the offspring will have the same beneficial traits as the original cow. It may also be difficult, given the prize-winning cow's advanced age, to extract enough eggs from her to carry out this technique (1 mark). For these reasons, the farmer has decided to use SCNT over embryo splitting.

You must include an outline of each process (2 marks). 'Discuss' means 'provide points for and/or against'. You must provide at least two points for and/or against each technique, and relate these points to the farmer's situation (4 marks).

Question 14

A benefit of the use of IVF in males with CF is that it provides them with an opportunity to have biological children, which for some men may be very important (1 mark). Even though the sperm of males with CF are healthy, they cannot fertilise the egg naturally because of the faulty or missing sperm ducts. IVF bypasses this problem by retrieving sperm directly from the testicles and using them to fertilise the egg outside the body (1 mark).

A benefit of the use of pre-implantation genetic diagnosis (PGD) in males with CF is that it allows embryos produced by IVF to be screened for CF, so that only those embryos that are free from CF are implanted. This ensures that the children of males with CF are free of the disease (1 mark). This would relieve the burden and stress of potentially passing on a fatal disease to their children (1 mark). If this technology were not available, some males with CF might choose not to have children, despite having the desire to do so.

> To score full marks, you must explain two benefits. 'Explain' means to link cause and effect. Identification of a benefit (e.g. it provides the opportunity for men with CF to have healthy children) is not sufficient – you must link the benefit (i.e. the effect) with the reason (i.e. the cause).

Question 15

a Restriction enzymes recognise a particular recognition site (1 mark) and cut the DNA, producing sticky ends (1 mark).

b DNA ligase seals the breaks in the sugar–phosphate backbone of DNA, sticking the inserted gene to the plasmid (1 mark).

c Some of the host bacteria could be grown in a broth containing ampicillin (1 mark). If the host bacteria had taken up the plasmid, whether it is recombinant or not, gene *A* will allow the bacteria to continue to multiply. This will show that transformation was successful (1 mark). If tetracycline is added to the broth and the bacteria survive, then the plasmid is not recombinant (1 mark). If the bacteria do not survive, then the recombination was successful, because the insulin gene would disrupt the function of gene *B* (1 mark).

> Note that you are asked to explain how to check for both recombination *and* transformation. You should address each of these separately.

Test 9: Causes of infectious disease

Multiple-choice solutions

Question 1

D Protozoa ingested in water, causing diarrhoea and vomiting

> Pathogens are organisms that cause disease in other organisms. **A** is incorrect because bacteria that live in the intestinal tract and enable digestion of food are beneficial to the host. **B** is incorrect because the fungi in the urogenital tract prevent infection by other pathogens. **C** is incorrect because the modified viruses do not cause disease, they destroy cancerous cells.

Question 2

C Exposure to UV inhibits reproduction of bacteria.

> As UV dosage increases, the number of bacteria present decreases, which suggests that exposure to UV radiation inhibits their growth. **A** is incorrect because no control plates are shown. **B** is incorrect because the highest dosage of UV shown did not eliminate all bacteria, and no data for higher UV dosages are shown. **D** is incorrect because it is UV radiation that reduces the amount of bacteria. The presence of bacteria cannot affect the amount of UV radiation.

Question 3

A UV dose

The scientists changed the dosage of UV radiation that each plate was exposed to. **B** is incorrect because the number of bacterial colonies counted is the dependent variable. **C** is incorrect because incubation temperature is a controlled variable. **D** is incorrect because inoculation with the same amount of water is a controlled variable.

Question 4

A An epidemic

The outbreak in Kismayo is an epidemic because it is more than the expected number of cases in that community in the time period. **B** is incorrect because the disease must have spread to several countries or continents for it to be considered a pandemic. **C** is incorrect because endemicity refers to the constant presence of a disease in an area at expected or baseline numbers. **D** is incorrect because an eradicated disease is one that has been permanently eliminated.

Question 5

D Bacterium

The diagram shows a cell wall, flagellum and DNA floating in the cytoplasm rather than in a membrane-bound nucleus. **A** is incorrect because viruses are non-cellular. **B** is incorrect because prions are non-cellular. **C** is incorrect because protozoa do not have a cell wall and have their DNA within a membrane-bound nucleus.

Question 6

B It is likely that syphilis is caused by the bacterium *Treponema pallidum*, but this has not been confirmed by Koch's postulates.

The presence of *Treponema pallidum* in every case of syphilis makes it likely that this bacterium is the causative pathogen, but the inability to grow it in pure culture means that not all criteria have been fulfilled. **A** is incorrect because all postulates should be fulfilled to state with certainty that a particular disease is caused by a particular pathogen. **C** is incorrect because, although the inability to grow the bacteria in culture prevents fulfilment of all Koch's postulates, this does not rule out the bacteria as a likely cause of the disease. **D** is incorrect because Koch's postulates are used for infectious disease.

Question 7

C Indirect and vector transmission

The diagram shows that cholera may be transmitted indirectly through fluids, food and so on, and also through flies, which act as vectors. **A** is incorrect because it only accounts for indirect transmission and not the vectors. **B** is incorrect because there is no direct contact between hosts. **D** is incorrect because although indirect transmission is occurring, there is no direct contact between hosts.

Question 8

C The nutrient broth was not boiled thoroughly on Day 1.

Both test tubes were cloudy after 14 days, indicating microbial contamination. This suggests that the broth was not boiled thoroughly, allowing some microbes to remain alive in the broth. **A** is incorrect because Pasteur used a nutrient broth and there is no evidence provided that suggests a different broth was used here. **B** is incorrect because broth does not always turn cloudy. This only happens in the presence of microbes. **D** is incorrect because exposure to oxygen does not cause cloudiness.

Question 9

B Susceptibility to antimicrobial substances

A pathogen that is susceptible to antimicrobial substances is less likely to be transmitted from one host to another. **A** is incorrect because the ability to survive outside the body in soil or water reservoirs increases the chance that the pathogen will be transmitted to a new host. **C** is incorrect because endospores prevent the pathogen from drying out, which allows it to live outside the body for some time. **D** is incorrect because remaining airborne for longer increases the chance that the pathogen will be inhaled by a new host.

Question 10

A An outbreak would result in financial loss through reduced egg production and deaths of birds.

Agricultural diseases that are not endemic to Australia can cause significant financial loss and death of animals. **B** is incorrect because there is no indication that humans would contract the disease by eating the infected chickens. **C** is incorrect because most farmers would be aware of the symptoms of any diseases that could affect their productivity. **D** is incorrect because the disease is caused by a virus and would not be treated with antibiotics.

Short-answer solutions

Question 11 (3 marks)

Pathogen	Cellular or non-cellular	Description of pathogen	Disease caused by this type of pathogen
Protozoa	Cellular	Single-celled, eukaryotic organisms with no cell wall	Malaria
Prion	Non-cellular	An abnormal protein that can cause diseases by making normal proteins fold incorrectly. Does not contain DNA or RNA	Creutzfeldt-Jakob disease
Virus	Non-cellular	Consists of a protein coat that encloses the DNA or RNA	Influenza A

There are many possible examples of diseases caused by these pathogens, as per previous notes.

Question 12

For example: One plant disease affecting the Australian agricultural industry is Panama disease in bananas, caused by a fungus called *Fusarium oxysporum*. The fungus causes the leaves to turn yellow and wither, and the stems to split. As the conducting tissues are damaged, the plant is starved of food and water and eventually dies. A second example is tomato bacterial wilt, caused by the bacterium *Ralstonia solanacaerum*. This bacterium causes damage to the roots, which prevents the plant from taking in water, causing the plant to wilt and die.

Agricultural crop industries in Australia contribute $29.3 billion to the economy each year. Plant diseases have the potential to reduce production in these industries as well as to threaten native plants, damaging Australia's unique environment as well as the forestry industry, which is worth a further $2.7 billion annually. This would have a significant impact on the economy of Australia and potentially lead to loss of employment for thousands of people. (5 marks)

The answer above is an example only. There are many ways to answer this question. As the question asks for named examples, it is necessary to include at least two. For each example, the name of the pathogen responsible for the disease (cause) must be included, along with the symptoms of the disease (effect). An assessment of these diseases must also be included, with a clear judgement of the extent of the impact of these diseases on agricultural production in Australia. As a stimulus is provided, it is essential to refer to it.

Question 13

a Louis Pasteur conducted an experiment in which he placed beef broth into swan-necked flasks that were unsealed. The broth was boiled to kill any microbes in the broth. Air was able to enter the flasks but any microbes in the air were caught in the curve of the swan-neck (1 mark).

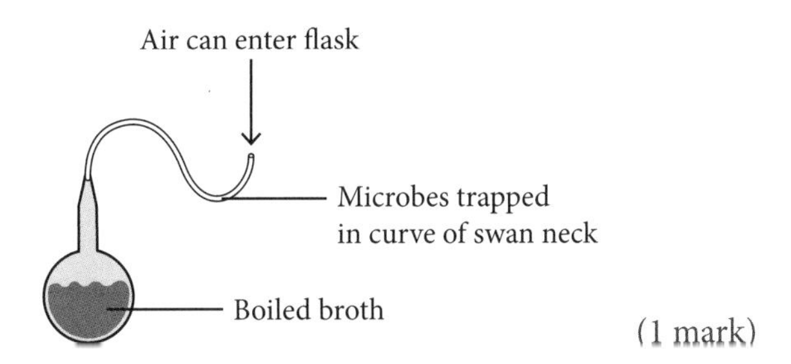

(1 mark)

Broth in these swan-necked flasks did not spoil (i.e. there was no bacterial or fungal growth). However, the broth spoiled when the swan-neck tubing was broken or when the flask was tipped, allowing the broth to reach the curve of the tube. This showed that it was microbes in air that caused spoilage of food, helping to refute the theory that living things could spontaneously appear (1 mark). Building on this work, Pasteur was able to show that disease was caused by the presence of a particular pathogen. He showed that diseases could be transmitted between organisms with no known contact, increasing our understanding of disease transmission. Pasteur's work also led to improvements in food preparation, hygiene, sterilisation and sanitation to prevent transmission of disease (1 mark).

b (2 marks)

Risk	Hazard	Mitigation measures
Pathogens could be inhaled or ingested	Pathogens could cause illness	Face masks can be worn to prevent inhalation. Gloves can be worn, and hands washed thoroughly after completing the investigation.
Sterilisation of the broth	Bunsen burner could cause burns or fire	Hair must be tied back and loose clothing secured. Flammable liquids used to sterilise bench (e.g. alcohol) must be kept away from the Bunsen flame.

Other answers are possible.

Question 14

a During the Olympic Games, athletes and officials from all over the world come together and then return to their home countries (1 mark). With an outbreak of Zika in Rio de Janeiro, there was the potential for many to become infected and then return home, spreading the disease all over the world (1 mark).

b Zika is spread by a mosquito vector (1 mark). Mosquitoes are more numerous in summer than in winter, so the games were held in winter to reduce transmission (1 mark).

c From the graph, it is apparent that both Zika infections and infant microcephaly increased over a period of three months before reaching a peak and then starting to decline (1 mark). These peaks occur approximately six months apart, indicating that the women infected with Zika early in their pregnancy were at increased risk of having a child born with microcephaly (1 mark). This suggests that Zika is not only transmitted by mosquito vectors, but from mother to child across the placenta (1 mark).

Question 15

Warren and Marshall's procedure was mostly reliable as they examined the stomachs of many patients with ulcers and gastritis, using an endoscope, which would allow them to observe the stomach lining clearly and take samples. The samples that they took were studied under the microscope, having been stained so that the bacteria could be observed and identified.

The procedure followed some of Koch's postulates for determining the cause of a disease, but it is not clear whether all were followed. For example, the first postulate is that the organism must be present in every case of the disease. It is implied that *Helicobacter pylori* is present in every case of the disease, but this is not explicitly stated.

The second postulate is that the bacterium must be isolated and grown in pure culture. The information provided does not clearly state that this occurred. This step is essential if it is to be conclusively shown that *H. pylori* causes stomach ulcers and gastritis.

The third postulate states that a potential host when inoculated with the bacterium should develop the same symptoms as the original host. Marshall inoculating himself with a dose of the bacteria, and developing the same symptoms, meets this postulate. However, a sample size much larger than one person is required in order for this to be reliable data.

The final postulate states that the microorganism must be isolated from the second host and identified as the same one found in the original host. The information does not state that Marshall and Warren carried out this final step.

Thus, while the methods used by Marshall and Warren used some appropriate technology and followed some appropriate procedures, the reliability and validity of the experiment could be improved in order to clearly show that *H. pylori* causes stomach ulcers. Reliability of the experiment would be improved by repeating the experiment with a much larger sample. Validity would be improved by ensuring that the pathogen was indeed present in every case of the disease and grown in pure culture, allowing them to be sure that it was the only pathogen introduced when inoculated into the healthy host. Validity would also be improved by re-isolating the bacterium and identifying it, showing that it had caused the symptoms shown by the second host. (8 marks)

There are different ways to approach this question. To gain full marks, a clear judgement must be made about the reliability, validity or accuracy of the procedures used (2 marks). The answer must also include reference to a variety of methods used (at least three) and at least one issue (4 marks). Reference to the methods and issues must be clearly supported with information from the stimulus provided (2 marks).

Test 10: Responses to pathogens

Multiple-choice solutions

Question 1

A A passive, physical barrier

The cuticle is a passive barrier, as it is always present in the plant and is not released in response to a pathogen's presence. It is a physical barrier as it prevents entry of the pathogen. **B** is incorrect because the cuticle is not a chemical released to reduce growth of the pathogen. **C** is incorrect because rapid active responses occur minutes to hours after invasion by a pathogen and are not always present in the plant. **D** is incorrect because delayed active responses occur in the days following invasion to prevent the spread of the pathogen.

Question 2

C Salicylic acid acts as a signalling agent for subsequent infections, limiting the severity.

Salicylic acid acts as a signalling agent for subsequent infections after an initial infection, playing a role in the plant's 'memory' of the pathogen. **A** is incorrect because the chemical receptors on the surface of plant cells are a chemical, passive response that occurs regardless of whether the pathogen has been encountered before. **B** is incorrect because leaves hanging vertically are a physical, passive response that occurs regardless of whether the pathogen has been encountered before. **D** is incorrect because small stomata are a physical, passive response by the plant regardless of whether the pathogen has been encountered before.

Question 3

B Apoptosis (programmed cell death) results in a cluster of dead cells surrounding the pathogen and isolating it.

Apoptosis is a response that can occur in both plant and animal cells. **A** is incorrect because animal cells do not have cell walls. **C** is incorrect because plants do not produce antibodies. **D** is incorrect because plants do not produce histamine.

Question 4

B The skin and chemical barriers

Skin is a physical barrier to prevent entry of pathogens if intact. There are antimicrobial substances on the skin that kill microbes or inhibit their reproduction. **A** is incorrect because phagocytosis occurs at the site of infection, after the pathogen has already entered the body. **C** is incorrect because both the inflammatory response and phagocytosis occur at the site of infection, after the pathogen has already entered the body. **D** is incorrect because the inflammatory response occurs at the site of infection, after the pathogen has already entered the body.

Question 5

C moving trapped pathogens to the mouth.

Cilia are small hair-like structures that move pathogens that have been trapped by mucus upwards towards the mouth to be expelled from the body. **A** is incorrect because cilia are inside the body, not outside. **B** is incorrect because cilia do not produce secretions. **D** is incorrect because cilia do not have a role in circulation of blood.

Question 6

D To trigger the dilation of blood vessels and increase their permeability

Histamine triggers the dilation of blood vessels and increases their permeability. **A** is incorrect because this describes the role of pyrogens. **B** is incorrect because histamine does not affect the pH of blood. **C** is incorrect because it is platelets that stop bleeding and re-establish the barriers.

Question 7

C	Neutrophil	Releases chemicals to inhibit pathogens Recruits other cells to site of infection	Migrates from blood vessels into tissue

Neutrophils release chemicals to inhibit pathogens and recruit other cells to the site of infection. They migrate from blood vessels into tissue. **A** is incorrect because mast cells are not antigen-presenting or involved with phagocytosis. They do not circulate in the blood. **B** is incorrect because macrophages do not cause blood vessel dilation or release histamine. **D** is incorrect because monocytes do not defend against parasites, release histamine or produce allergic reactions. They are stored in the spleen.

Question 8

D Pyrogens

A is incorrect because interleukins are chemical messengers that link the innate and adaptive immune systems. **B** is incorrect because cytokines are chemical messengers that promote the development and differentiation of B and T cells. **C** is incorrect because monocytes are cells involved with phagocytosis.

Question 9

A Mucous membranes, tight junctions between cells, sphincter muscles in the urethra

Mucous membranes, tight junctions between cells and sphincter muscles in the urethra are all physical barriers. **B** is incorrect because oil and tears are chemical barriers. **C** is incorrect because gastric secretions are a chemical barrier. **D** is incorrect because enzymes, fatty acids and oil are chemical barriers.

Question 10

C A physical response, as urination flushes pathogens out of the body.

Urination flushes pathogens out of the body, so this is a physical response. **A** is incorrect because it is not a chemical response and more urea is not produced. **B** is incorrect because urination does not produce more chemicals. **D** is incorrect because the urine does not contain phagocytes.

Short-answer solutions

Question 11

a For example: Plants in the Myrtaceae family (e.g. *Callistemon* and *Melaleuca*) may develop a condition called myrtle rust when infected with the fungal pathogen *Austropuccinia psidii* (1 mark). Myrtle rust causes deformed leaves, stunted growth and possibly death of the plant. Plants exhibit many defences against these conditions, including a thick cuticle acting as a physical barrier, chemical barriers (e.g. saponin glycosides), releasing hydrogen peroxide as a rapid response and antimicrobial substances to limit the spread (1 mark).

There are many examples that could be used to answer this question. For full marks, the answer must include the name of a plant or plants affected, as well as the pathogen that causes the disease, then outline at least one response that the plant exhibits to respond to an invasion by the pathogen.

b *Aim*

To investigate the response of *Callistemon* plants to the fungal pathogen *Austropuccinia psidii*

Materials

- Samples of diseased *Callistemon* plants
- Samples of healthy *Callistemon* plants
- Dissecting microscope
- Forceps

Method

1. A dissecting microscope was set up.
2. A sample of a diseased *Callistemon* plant was examined under the dissecting microscope and signs of disease were observed and recorded using diagrams and photographs.
3. A sample of a healthy *Callistemon* plant was examined under the dissecting microscope for comparison and the results were recorded using diagrams and photographs.
4. The samples of both diseased and healthy plants were also examined for responses to the pathogen, including thick cuticle and release of enzymes and chemical substances.

Risk assessment

- Gloves should be worn to prevent allergic reactions to the leaves.
- A face mask should be worn to prevent inhalation of fungal spores from infected plants. (4 marks)

The answer above is an example. The question could be approached in a variety of ways. For full marks, the response must include:

- all materials required (1 mark)
- a detailed step-by-step method that describes how the diseased samples were examined and compared to healthy plants (2 marks)
- a risk assessment that includes at least one valid risk and mitigation strategy associated with examining diseased plant samples (1 mark).

9780170465250

Question 12 ©NESA 2014 MARKING GUIDELINES SII Q21 (ADAPTED)

a Phagocytosis (1 mark)

b Increased blood flow to the site of infection (1 mark) brings more white blood cells to engulf the pathogen (1 mark).

c Macrophages (1 mark)

> Many answers would be accepted for part c, including monocytes, phagocytes, basophils and neutrophils.

Question 13

a The gastric secretions in the stomach contain hydrochloric acid, creating a highly acidic environment with a pH of 1–2 (1 mark). The acidic environment discourages growth of pathogens, and most will not survive (1 mark) – thus the host is protected against infection.

b Tight junctions between endothelial cells (1 mark) prevent the movement of pathogens from infected tissue into the blood vessels (1 mark).

> Many answers would be accepted for part b. Other examples are mucous membranes, peristalsis, sphincter muscles, mucus, cilia, vomiting.

Question 14

a

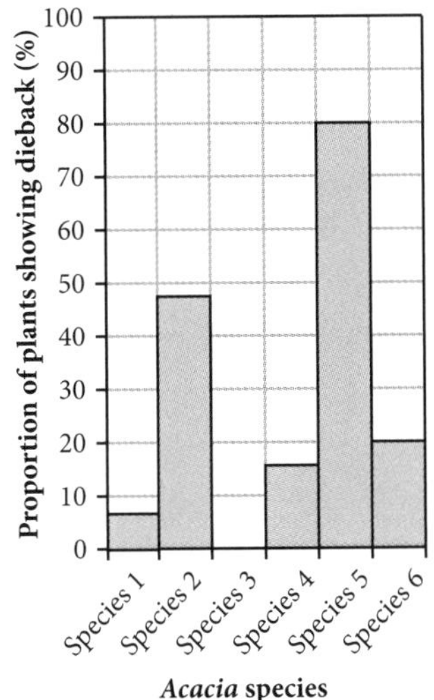

> A column graph is most appropriate because the data are discrete rather than continuous. The graph should include the following:
>
> - both axes correctly labelled, including units for *y*-axis
> - a correct and even scale
> - data neatly and accurately plotted on the scale
> - a correct and informative title.
>
> For full marks, all of the above must be included. For 2 marks, one of the elements may be absent. 1 mark may be allocated for some relevant information provided, with two or more of the elements missing. Note that if a line graph is provided instead of a column graph, a maximum of 1 mark would be given.

b The *P. cinnamomi* pathogen causes damage to the roots of plants. Releasing an enzyme that increases lignin production and strengthens the cell walls of root cells would prevent the pathogen from damaging the roots (1 mark), so they would be able to perform their primary function of absorbing water from the soil, preventing the plant from dehydrating and dying (1 mark). From the graph, it is most likely that *Acacia* species 3 has developed this adaptation because 0% of the population showed signs of dieback (1 mark).

Question 15

The skin covers the whole body and acts as a physical barrier to prevent entry of pathogens. As long as the skin is intact, most pathogens are unable to penetrate it and gain entry to the body in this way. Skin has a layer of keratin, which is waterproof and tough, providing resistance to enzymes secreted by pathogens.

The skin also provides a chemical barrier, secreting antimicrobial substances that kill pathogens on the surface of the skin.

When an injury occurs, such as that shown in the diagram, the integrity of the skin as a barrier is compromised. The layer is no longer intact, so pathogens could enter through the opening created. The splinter that pierced the skin is also likely to have had microbes on its surface, which have now been introduced into the body. A number of physical and chemical responses would occur in the woman's body to prevent the spread of infection.

Physical responses include:

- cell death to seal off an area and trap the pathogen: if infected cells are surrounded by a wall of dead cells, the infection is unable to spread to healthy cells. This wall of dead cells forms a capsule called a granuloma. The contents of the granuloma will be broken down and removed by phagocytosis
- vasoconstriction, which prevents bleeding; and platelets stick to the wound to form a plug and clot the wound. This prevents entry of any more microbes to the body via this wound.

Chemical responses include:

- the inflammatory response, in which the cells release chemical 'alarm signals' (e.g. histamine), which cause vasodilation and increase the permeability of blood vessels, allowing phagocytes to move from the bloodstream into the tissues to the site of infection. This allows phagocytosis to occur
- release of pyrogens, which cause the body temperature set point to be raised beyond the normal temperature of 37°C (fever). This kills or limits the growth of pathogens and may increase the activity of white blood cells
- Release of cytokines, which stimulate the development of B cells and T cells and initiate the adaptive immune response. (8 marks)

This question can be approached in different ways. Many different physical and chemical responses could be included. For this question, marks would be allocated as follows:

- a description of the role of the skin, including its role in providing a physical and chemical barrier to preventing entry of pathogens if intact – 2 marks
- physical responses – 3 marks. At least two examples should be included. The question states that the answer should *account* for these responses, so for full marks, an explanation of how they prevent the spread of infection should be included.
- chemical responses – 3 marks. As above, at least two examples should be included, with an explanation of how they prevent the spread of infection.

Test 11: Immunity

Multiple-choice solutions

Question 1

B a molecule that triggers an immune response in the host.

Antigens are molecules that trigger an immune response in the host. **A** is incorrect because this is the definition of antibiotics. **C** is incorrect because this is the definition of an antibody. **D** is incorrect because antigens are not involved in production of lymphocytes.

Question 2

C

Characteristic	T cell	B cell
Cell surface receptor can recognise a specific antigen	✓	✓

Both B and T cells have cell surface receptors that recognise specific antigens. **A** is incorrect because T cells do not produce plasma cells. **B** is incorrect because T cells are not involved in antibody production. Although B cells are involved in antibody production, they are not released in body fluids. **D** is incorrect because both B and T cells form clones once stimulated.

Question 3

B The adaptive response is specific to a pathogen and has 'memory' to improve the immune response to pathogens on future exposure.

A is incorrect because the innate response is not specific to a pathogen and does not have memory to improve the immune response for consequent infections. **C** is incorrect because no memory cells are produced in the innate response. **D** is incorrect because the adaptive response is specific to a pathogen.

Question 4

D

	Inside cells	Outside cells
Protozoa	✓	×
Bacteria	✓	×
Virus	✓	×
Fungi	×	✓

A is incorrect because the cell-mediated response does occur inside the cells for bacteria and does not occur outside cells for viruses. **B** is incorrect because the cell-mediated response does not occur inside cells for fungi, only outside cells. **C** is incorrect because the cell-mediated response occurs only inside cells for protozoa, bacteria and viruses and only outside cells for fungi.

Question 5

B This is programmed cell death and it is part of the both the innate and adaptive immune responses.

A is incorrect because programmed cell death is also part of the innate response. **C** and **D** are incorrect because the diagram does not show the humoral response.

Question 6

C MHC molecules

A is incorrect because cytokines stimulate production of B and T cells; they are not involved in recognition of self and non-self. **B** is incorrect because interleukins are a type of cytokine. **D** is incorrect because antibodies are proteins produced to counteract a specific antigen.

Question 7

B II and III

All other options include statements I or IV, which contain incorrect information.

Question 8

A Cytotoxic T cells migrate to the site of an infection, and their antigen receptors bind to the antigen displayed on their surface.

All other options are incorrect because they include innate responses.

Question 9

C The innate immune system is fully functional, and the baby may have some temporary antibodies received from the mother through the placenta and breastfeeding.

The innate immune system is what a person is born with. Some temporary antibodies produced by the mother may be present, via the placenta and through breast milk. **A** is incorrect because the adaptive immune response is not fully functional at birth. **B** is incorrect because the baby is likely to have some temporary antibodies received from the mother via the placenta and/or breast milk. **D** is incorrect because the baby's adaptive immune system is not fully functional at birth, with any antibodies received from the mother being only temporary. It does not take time for the innate system to develop – it is functioning at birth.

Question 10

C Stimulation of a B cell to become a plasma cell

T cells release cytokines, which stimulate B cells to differentiate to form plasma cells. **A** is incorrect because the antigen is already displayed on the surface of the B cell, meaning this occurred previously – the diagram is focused on the interaction between the B cell and the T cell. **B** is incorrect because the diagram does not show a macrophage. **D** is incorrect because cytotoxic T cells puncture the cell and inject chemicals to destroy the cell, which is not happening in this diagram.

Short-answer solutions

Question 11

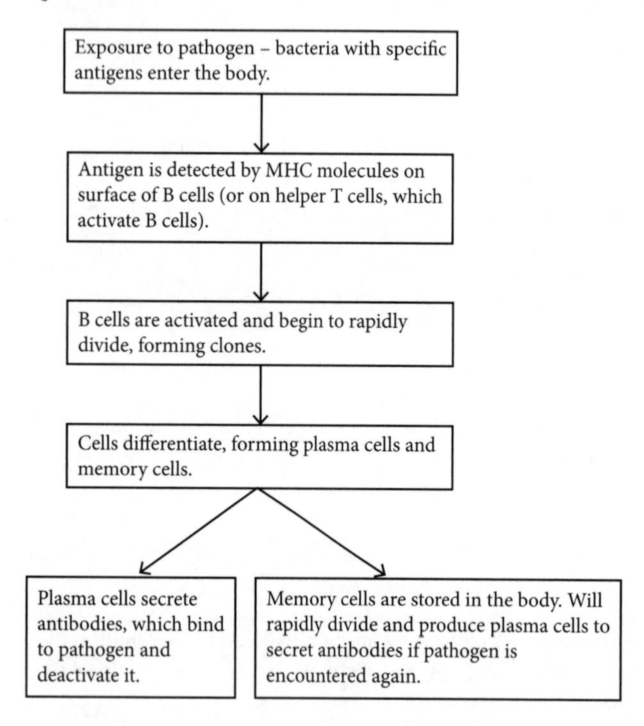

This question may be approached in different ways, with different information included at each step. For full marks, the information must be presented as a flow chart, as stated in the question, and the following must be included:

- specific detail on how the pathogen is detected and how the B cells are activated
- reference to differentiation of cells to form plasma cells as well as memory cells
- reference to the secretion of antibodies by plasma cells
- reference to the storage of memory cells to protect against subsequent infection.

Question 12 ©NESA 2010 SAMPLE ANSWERS SII Q28 (ADAPTED)

a An antigen (1 mark)

b The model is highly effective, as it shows the difference between the two types of cells (1 mark). The two cells have different surface molecules (antigens) (1 mark), which means the recipient's immune system will be able to recognise and attack the new organ (1 mark).

This answer can be approached in different ways. For full marks, the answer must include a judgement of the effectiveness of the model, supported with reference to a feature of the model provided, and explaining how the feature aids understanding of organ rejection.

c Models are used to represent complex concepts that are difficult to observe directly, enabling the concept to be explained and increasing understanding (1 mark). Using simple symbols to show that the cells have different antigens on the surface helps to show why transplanted organs may be rejected by the body (1 mark).

d T killer cells (1 mark) recognise and bind to transplanted cells and release factors that kill them (1 mark). T helper cells (1 mark) produce growth factors that help other lymphatic cells (1 mark). For example, B cells grow and produce factors (antibodies) that cause the death of transplanted cells.

Question 13

a Phagocytosis is the process by which phagocytes (e.g. macrophages and neutrophils) engulf an invading pathogen and destroy it with the use of enzymes. (3 marks)

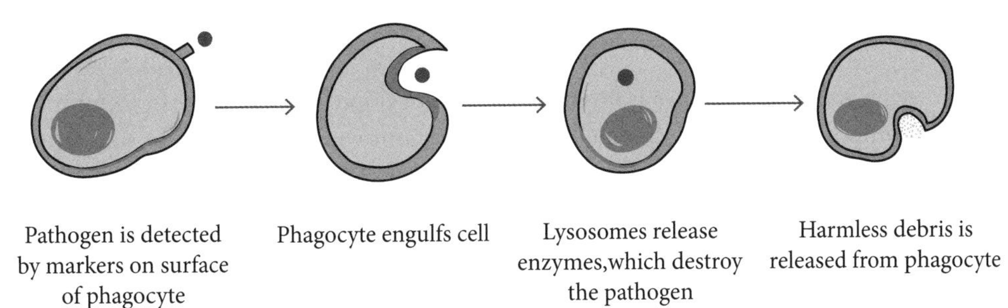

Pathogen is detected by markers on surface of phagocyte

Phagocyte engulfs cell

Lysosomes release enzymes,which destroy the pathogen

Harmless debris is released from phagocyte

This question can be approached in different ways. It is essential that a simple diagram is included along with all the information shown in the annotations. However, it would be acceptable to include some of the detail in a written explanation, using the diagram to support the response.

b Phagocytosis is part of both the innate and adaptive immune responses. It is a non-specific response that occurs quickly after invasion by a pathogen and is thus part of the innate response (1 mark), but it is also responsible for 'cleaning up' pathogens and debris after inactivation by antibodies or cytotoxic T cells and is thus part of the adaptive immune response (1 mark).

Question 14

a

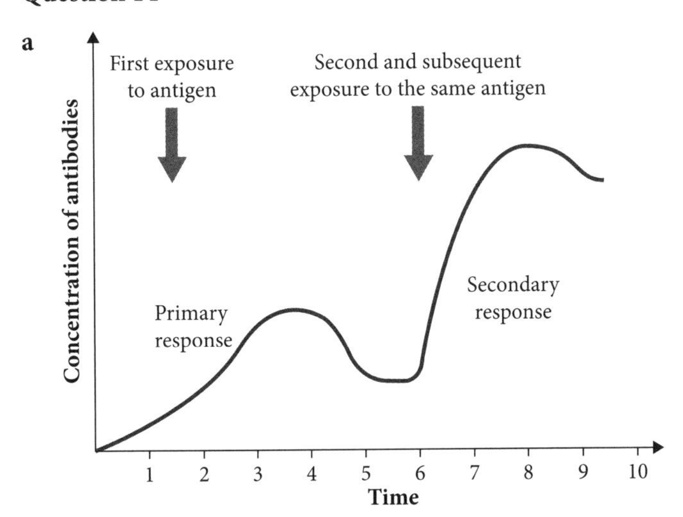

The response must show the initial concentration of antibodies in the blood as 0, then rising gradually before declining, but not returning to 0 (1 mark). The secondary response must be shown to be much more rapid and reaching a higher concentration (1 mark).

b The primary response to a pathogen is production of antibodies, but this response is short-lived. When exposed a second time, the production of antibodies is much greater and more rapid (1 mark), resulting in long-term immunity against the disease. In receiving a second dose of a vaccine, the individual experiences a secondary exposure and develops long-term immunity against the disease (1 mark).

Question 15

The adaptive immune response provides a specific response to pathogens, deactivating them so that they no longer cause harm to the host. It is also through the adaptive immune response that memory cells are made, and these can remain in the body, mounting a rapid response if the host is reinfected with the same pathogen again. Without the adaptive immune response, a host could not develop long-term immunity. (1 mark)

However, the innate immune response also provides essential responses to a pathogen. For example, many pathogens are unable to enter the host's body at all, because of physical and chemical barriers that form a part of the innate immune response. If these are successful, there is no need for the adaptive immune response (1 mark).

Additionally, non-specific processes in the innate immune response (e.g. inflammation, phagocytosis and cell death) can prevent infection from spreading rapidly through the body, allowing time for the adaptive immune response to occur before the host is overwhelmed by the pathogen (1 mark).

Thus, although the adaptive immune response is essential for the development of long-term immunity, the processes of the innate response are also essential, with the two systems working together (1 mark).

This question can be answered in different ways. The response should include arguments for and against the statement – that is, benefits of the adaptive immune response as well as benefits of the innate immune response. A judgement of the validity of the statement should be made, with reference to the essential nature of both responses.

Test 12: Prevention, treatment and control

Multiple-choice solutions

Question 1

D Penicillin may have played a role in the survival of the four treated mice.

As all mice that had penicillin survived, and all that did not receive penicillin did not survive, penicillin may have had a role in the survival of the treated mice. **A** is incorrect because although more mice would increase the reliability of the data, there is enough evidence that the penicillin may have had an effect to draw this conclusion. **B** is incorrect because there is no evidence of antibiotic resistance; the mice that received antibiotic survived and those that did not receive it died. **C** is incorrect because the antibiotic was only used on mice, so it might not be safe for use in humans.

Question 2

D Ban the import of live animals, uncooked meat and unprocessed dairy into Australia from countries with FMD.

Banning the import of live animals, uncooked meat and unprocessed dairy into Australia from countries with FMD could prevent entry of FMD into Australia. **A** is incorrect because treating animals on Australian farms would not prevent entry of the disease into Australia; it would already have entered Australia. **B** is incorrect because animals may not exhibit symptoms prior to leaving their country of origin but may develop symptoms upon arrival in Australia, and thus this method does not prevent entry of the disease into Australia. It would also rely on farmers in other nations reporting symptoms. Also, antibiotics are not effective against viral diseases, so treating the cattle with antibiotics before export would not help to prevent FMD from entering Australia. **C** is incorrect because wearing masks and gloves and washing hands would not prevent entry of the disease into Australia if the animals were imported.

Question 3

D The message is clear, simple and easy to remember, so it is more likely to be followed.

A is incorrect because there is no guarantee that a person won't get the virus if they follow these guidelines, even though it is likely to help reduce the transmission of the disease. **B** is incorrect because these steps may reduce the chance of transmission but will have no effect on the severity if transmission occurs. **C** is incorrect because the instructions are not detailed, they are very simple.

Question 4

A 1.35%

1102 people out of a 'population' of 81 510 had Spanish influenza, so the percentage is $\frac{1102}{81510} \times 100 = 1.35\%$.
All other answers are incorrect because they have not been obtained by taking the number of cases and dividing it by the population size.

Question 5

B These measures would slow the spread of the disease, by minimising contact between individuals and preventing transmission via airborne droplets.

A is incorrect because no method is completely effective at preventing transmission of a disease. **C** is incorrect because the vaccine that was eventually produced did not prevent the spread of the disease, it simply reduced the chance of succumbing to a secondary infection, so these additional measures were still necessary. **D** is incorrect because Spanish influenza was highly transmissible.

Question 6

B Spraying crops with pesticides to kill pests

Spraying crops with pesticides to kill pests would not prevent viral diseases in plants. **A** is incorrect because genetic engineering can be used to give plants natural resistance to viruses. **C** is incorrect because preventing the importation of live plant materials into Australia could protect agricultural plants from imported viruses. **D** is incorrect because burning seized infected plants will destroy the virus and prevent it from entering Australia.

Question 7

A Increased rainfall leading to a malaria outbreak in a number of equatorial countries

Increased rainfall leading to a malaria outbreak in a number of equatorial countries occurs at a regional level. **B** is incorrect because transmission of a disease from animals to humans in a town is a local factor. **C** is incorrect because mobility of refugees who have received inadequate health care prior to migration is a global factor. **D** is incorrect because screening for tuberculosis by conducting chest X-rays on migrants before they migrate to a new country is a global factor.

Question 8

C The independent variable is the quarantine policy.

The independent variable is the factor that varies between groups. Each island has its own quarantine policy. **A** is incorrect because the smallest island has its own quarantine policy and so is one of the experimental groups. **B** is incorrect because the number of infected cattle is the dependent variable, not the control. **D** is incorrect because the number of infected cattle is the dependent variable, not the independent variable.

Question 9

B Educating the public so that they understand the risk of handing the rodents and abandon harmful practices

Educating the public to understand the risks of a disease and abandon harmful practices is likely to have the biggest impact on controlling spread of a disease in a population influenced by cultural practices. **A** is incorrect because setting traps for rodents is not likely to reduce spread of the disease when they are being used for preparation of charms and curses and thus deliberately handled. **C** is incorrect because it is not a viable solution to genetically engineer all rodents in the area so that they can't spread the disease. **D** is incorrect because it is unlikely that spiritual healers will change their behaviour in response to intervention by Western medical advice. Hygiene methods are unlikely to prevent transmission if rodents are still used.

Question 10

C There has been widespread distribution of mosquito nets and effective public education campaigns.

Widespread distribution of mosquito nets and effective public education campaigns could reduce the chance of being bitten by a mosquito, thus reducing the chance of transmission. **A** is incorrect because improving the conditions for mosquito vectors would increase the chance of transmission, which is not what the graph shows. **B** is incorrect because malaria is not caused by a virus. **D** is incorrect because people with malaria are not quarantined, because this measure is ineffective.

Short-answer solutions

Question 11

a Passengers and crew could be isolated to their rooms to prevent the spread of disease from infected people to healthy people (1 mark). Masks could be worn when individuals leave their rooms to prevent transmission by droplets in the air (1 mark). Improved hand hygiene may help to prevent transmission via surfaces such as handrails, doorknobs and so on (1 mark).

There are different possible answers to this question. This response is just an example.

b Because passengers were allowed to travel after disembarking from the *Ruby Princess*, the disease was spread to other states and potentially to other countries (1 mark). Other people who passengers came into contact with may have become infected, such as fellow travellers on planes or people in the communities they returned to (1 mark). Precautions that could have been used to prevent this include testing passengers before they disembarked to ensure that they did not have the disease, or a period of quarantine in Sydney before they were allowed to travel home (1 mark).

Question 12

An example of a plant used by Indigenous Australians is smokebush. Smokebush was used for wound healing and to treat pain (e.g. headaches) and inflammation (1 mark). Western pharmaceutical companies have been investigating smokebush for possible use in treating cancer and HIV/AIDS (1 mark).

In Australia, intellectual property laws currently allow anyone to apply for the rights to the knowledge and use of any native plant. This does not align with Indigenous cultural views that no one can have exclusive ownership of the land and that it is there for use by those who need it (1 mark). Current laws may also prevent Indigenous people from using resources that they have been using for thousands of years. It is essential to consider the customs of Indigenous people, consult with Elders and not allow individuals or companies to have exclusive rights to resources that are used by Indigenous people (1 mark).

This question can be answered in different ways. It is essential that a *named* example is used. Describing how this plant was traditionally used by Indigenous Australians and how it may be used today will earn 2 marks. Discussing the importance of respecting Indigenous culture and intellectual property will earn 2 marks.

Question 13

Antiretroviral therapy (ART) involves the administration of antiretroviral drugs that suppress the viral load of the HIV virus. This prevents the virus from hijacking CD4 cells, which are involved in protecting the body against other diseases.

The number of new cases of HIV was already decreasing prior to the introduction of ART in 2004, from a peak of approximately 3.3 million in 1997 to approximately 2.7 million in 2004. This is most likely due to education campaigns and availability of preventative measures such as condoms, needle exchange programs and blood screening for transfusions. The number of new infections has continued to decline since 2004 (with just under 2 million cases in 2019), suggesting that ART is effective at preventing new infections.

The number of deaths from HIV can be seen to decrease since the introduction of ART in 2004, from a peak of approximately 1.8 million in 2004 to approximately 850 000 in 2019. This suggests that ART is effective at suppressing the viral load and keeping the CD4 cells at normal levels.

The number of people living with HIV has increased since the introduction of ART, from approximately 27 million in 2004 to approximately 38 million in 2019. This does not mean that the treatment is ineffective but rather that people with HIV infection are now living with the disease rather than dying from it. Over time, as the number of new infections continues to decrease, it is expected that this number may begin to decline as well.

Thus, it can be seen from the information in these graphs that ART has been highly effective at reducing the number of infections and number of deaths from HIV, allowing people to live with the disease.

This question requires you to refer to all three lines on the graph, indicating how each factor has changed since the introduction of ART and stating whether ART appears to have had an impact on this factor (3 marks). Clear and thorough explanations with specific reference to data in the graph should be included (1 mark). A clear judgement of the effectiveness of ART treatment should be made (1 mark).

Question 14

a An example of an environmental management strategy would be to remove stagnant water reservoirs from the area (e.g. swamps, ponds), which would otherwise provide breeding places for mosquitoes (1 mark). If there are fewer suitable places for breeding, the numbers of the mosquitoes will be reduced (1 mark). The dengue and Ross River viruses are transmitted by mosquito vectors, so reducing the number of mosquitoes would reduce the spread of these diseases from one host to another (1 mark).

This question can be answered in many ways. The response must include a specific environmental management strategy, how this would reduce number of mosquitoes, and a link to reduction in transmission by reducing the number of mosquito vectors.

b With current climate change projections, it is expected that the *Aedes* mosquito will spread to other areas. It is currently only found in areas of northern Australia, where the population is quite small. If it was to spread to other areas of Australia with larger populations (e.g. capital cities), there is a risk of Ross River virus spreading to a large number of people (1 mark). This would be costly to the government in terms of providing health care and lost revenue if people are unable to work when they are sick (1 mark). Thus, developing and administering a vaccine may cost less, and it may be worth reconsidering the decision not to manufacture it (1 mark).

Question 15 ©NESA 2021 SAMPLE ANSWERS SII Q30

The measles vaccine protects children against contracting measles. The graph shows that the incidence of measles in vaccinated children is very low, sometimes zero, compared with that in unvaccinated children. Because these children are matched for age and other social factors, it is likely that the vaccine is responsible for the differences. The table supports this conclusion in that only two vaccinated children died of measles, compared with 40 in the unvaccinated group. Therefore the data supports the statement that the measles vaccine protected the community against measles. The table provides data that suggests the measles vaccine protects children against dying from diseases other than measles, and therefore contradicts the statement that a vaccine targets a specific disease. Children vaccinated against measles died from diarrhoea and dysentery at about half the rate of unvaccinated children, and less than one-third the rate from oedema. However, numbers dying of oedema are small, so this conclusion needs further study. If all other variables were taken into account, then the vaccine appears to have non-specific beneficial effects, especially for diarrhoea. However, the small differences between the groups with respect to fever suggest that the protection of the vaccine is for selected diseases only. Because it is likely that the vaccine targeted specific antigens for the measles virus, the mechanism for this wider protection is not clear. Overall, the vaccinated group showed about half the mortality of the unvaccinated group, supporting the conclusion that the vaccine was protective in a more general way. The results in the table raise questions about the specificity of the measles vaccine and therefore the statement requires qualification.

This question can be approached in different ways. The response should include an analysis and interpretation of the scientific data provided, as well as arguments for and against the statement.

If the analysis includes arguments only for *or* against the statement, a maximum of 5 marks would be awarded. An understanding of the role of vaccines in controlling infectious disease should be included.

Test 13: Homeostasis

Multiple-choice solutions

Question 1

D Horses sweat to lose heat by evaporative cooling.

Production of sweat is a physiological adaptation that allows heat to be lost by evaporative cooling. **A** is incorrect because fur is a structural adaptation. **B** and **C** are incorrect because the behaviours described are behavioural adaptations.

Question 2

A The ears have a large surface area to volume ratio, which would increase heat loss, so it probably lives in a hot climate.

B is incorrect because ears with a large surface area to volume ratio lose more heat, so the animal would not live in a cold climate. **C** is incorrect because large, flat ears have a large surface area to volume ratio. **D** is incorrect because large, flat ears have a large surface area to volume ratio.

Question 3

A Graph A

As the amount of water in the soil increases, the width of the stomatal opening increases, because there is more water available to absorb via the roots, so there is less need to prevent water loss through the stomata and they can remain open. **B** is incorrect because stomata close when the amount of water available is reduced, but this graph shows the opposite. **C** is incorrect because this graph does not show any relationship between water availability and the width of stomatal openings. **D** is incorrect because it does not show a clear relationship between water availability and the width of stomatal openings.

Question 4

D When salt levels in the body are too high, aldosterone is released to bring the levels back to the set point.

All other options are incorrect because they describe examples of positive feedback.

Question 5

B The brain

A is incorrect because the heart is not involved in the negative feedback response. **C** is incorrect because thermoreceptors pass information from one spot in the body to the brain, and the hypothalamus detects that the body temperature is below normal. **D** is incorrect because pressure receptors are unable to determine that the core body temperature is below normal.

Question 6

B Endocrine system

A is incorrect because the lymphatic system removes excess fluid from the body, absorbs fats and produces immune cells. **C** is incorrect because the integumentary system produces the protective layer of the skin and helps regulate body temperature. **D** is incorrect because the nervous system is responsible for transmitting electrical messages.

Question 7

C

Receptor	Stimulus	Effector	Response
Beta cells	Blood glucose increases	Body cells	Take up glucose

A is incorrect because the receptor, stimulus and effector have been incorrectly identified. **B** and **D** are incorrect because all components of the response have been incorrectly identified.

Question 8

D Glucagon

A is incorrect because insulin is released in response to an increase in blood sugar level. **B** is incorrect because glucose is a nutrient, not a hormone. **C** is incorrect because glycogen is the stored form of glucose.

Question 9

Graph A

A person with diabetes would have an elevated blood sugar level to start with, which would rise much higher than that of a person without diabetes and remain high for a longer period of time. **B** is incorrect because it shows the pattern reversed for the diabetic and non-diabetic individuals. **C** is incorrect because it indicates that there will be no change in blood sugar for a non-diabetic person and that the blood sugar of the diabetic person will continue to rise. **D** is incorrect because it indicates that there will be no change in blood sugar for a diabetic person and that the blood sugar will continue to increase for a longer period of time for a non-diabetic person.

Question 10

C

(i)	(ii)	(iii)
Sensory neuron	Interneuron	Motor neuron

All other options are incorrect because some or all of the neurons have been incorrectly identified.

Short-answer solutions

Question 11

a (3 marks)

Relationship between water temperature and blubber

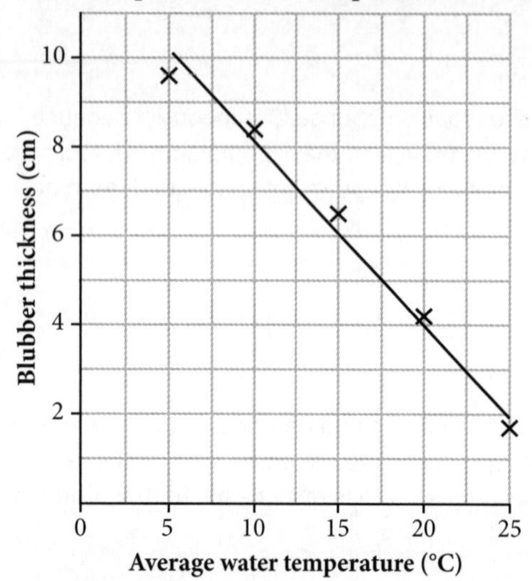

Responses require a title, labelled axes with units, an even scale, accurately plotted points and a line of best fit.

b As the water temperature increases, the blubber thickness of the sea lions decreases (1 mark). For example, at 5°C the average thickness is 9.6 cm, while at 25°C the average thickness is only 1.7 cm (1 mark).

> 1 mark is given for identifying the relationship between water temperature and blubber thickness. The other mark is given for making specific reference to data in the graph.

c Structural (1 mark)

Question 12

a The nervous system provides the neural pathways by which messages travel through the body (1 mark). This allows for coordination of activities that maintain homeostasis (1 mark).

b An action potential is an 'all-or-nothing' reaction: if the membrane potential reaches the threshold of −55 mV, the reaction will occur; if it does not reach this value, no action potential will occur (1 mark). The graph shows that stimulus A passes the threshold value of −55 mV but stimulus B only changes from the resting potential value of −70 mV to −65 mV (1 mark). Thus, stimulus A would result in an action potential, but stimulus B would not (1 mark).

Question 13

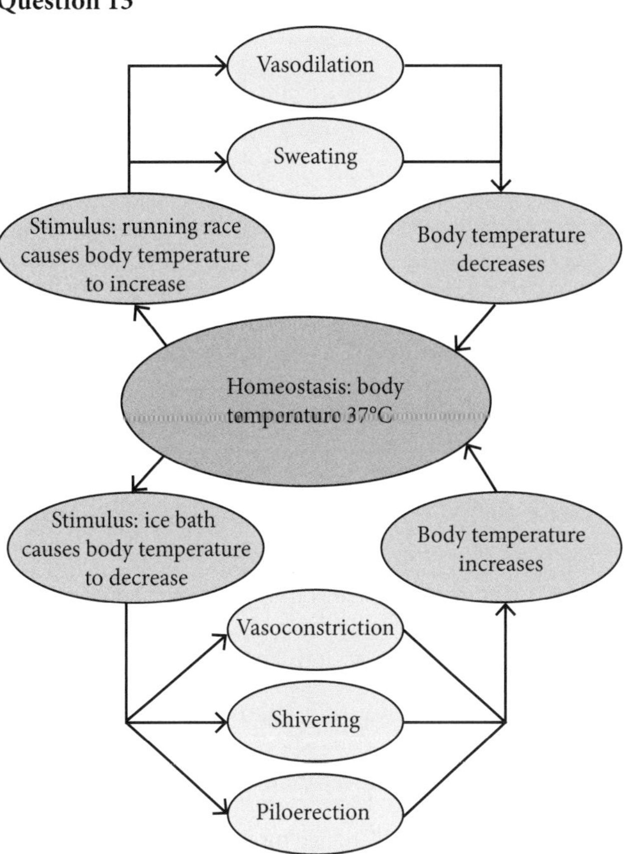

For this 6-mark question, 3 marks would be allocated to the negative feedback mechanism experienced in response to increased temperature caused by running, and 3 marks would be allocated to the negative feedback mechanism experienced in response to decreased temperature caused by the ice bath. The diagram should be set out as a loop rather than a flow chart, to show the return to homeostasis. Each side of the loop should include the stimulus, responses and effect of these responses.

Question 14

Banksias have woody fruits rather than fleshy fruits (1 mark). This prevents loss of water when the fruits fall to the ground (1 mark). Eucalypts have leaves that hang vertically (1 mark). This reduces exposure to direct sunlight, which would cause loss of water by evaporation (1 mark).

There are many possible answers to this question. To gain full marks, two examples must be provided and clearly explained. For each explanation, the mechanism must be clearly described (cause), along with a clear indication of how this prevents water loss (effect).

Question 15 ©NESA 2021 SAMPLE ANSWERS SII Q32

In a normal male, as shown in the diagram, the hypothalamus releases releasing hormone, which stimulates the anterior pituitary to produce hormones that stimulate the testes to produce sperm and the hormones testosterone and inhibin. Testosterone stimulates sperm production also. When a large dose of an anabolic steroid that is like testosterone is taken, while this may initially increase sperm production and the associated hormone inhibin, the testosterone effect of negative feedback on both the hypothalamus and the anterior pituitary cause less stimulatory hormones to be released, so the testes produce less sperm and less inhibin and less testosterone. Although less inhibin would mean a smaller negative feedback on the anterior pituitary caused by this hormone, the ongoing dose of testosterone would ensure a large overall negative feedback effect. In time, the testes would lose their normal function. (5 marks)

For full marks, the response must show a comprehensive knowledge of negative feedback loops and a correct interpretation of the diagram provided. A clear explanation (showing cause and effect) of the changes that occur in the testes in response to taking steroids must also be provided.

Test 14: Causes and effects

Multiple-choice solutions

Question 1

D It is caused by a nutritional deficiency.

Scurvy is a nutritional disease caused by lack of vitamin C, which is why the sailors who were given lemons and oranges recovered. **A** is incorrect because scurvy is not caused by sun exposure. **B** is incorrect because there is no gene for scurvy and no evidence that the disease was inherited. **C** is incorrect because scurvy is not passed on from rats (and those who ate rats did not seem to develop scurvy, which shows that eating rats was a protective measure, rather than causing scurvy).

Question 2

C Nutritional deficiencies

Nutritional deficiencies are not known to cause cancer. **A** is incorrect because some viruses can cause changes to DNA, resulting in cancer. **B** is incorrect because particular genes can increase the likelihood of developing cancer. **D** is incorrect because spontaneous, uncontrolled cell division can lead to cancer.

Question 3

B Non-Indigenous Australians are more likely to die from colorectal cancer than Indigenous Australians.

The graphs show that non-Indigenous Australians have a higher age-standardised mortality rate for colorectal cancer, indicating that they are more likely to die from it than Indigenous Australians. **A** is incorrect because Indigenous Australians are not more likely to develop colorectal cancer (as shown by the lower incidence rate). **C** is incorrect because non-Indigenous Australians are not more likely to survive colorectal cancer than Indigenous Australians, as shown by the higher mortality rate. **D** is incorrect because the chance of surviving colorectal cancer is lower in non-Indigenous Australians.

Question 4

D mortality rate.

A is incorrect because the incidence rate is the number of new cases of a disease in a specified time period. **B** is incorrect because the prevalence rate is the number of existing cases of a particular disease at a particular point in time. **C** is incorrect because the morbidity rate is the percentage of the population who develop a particular disease.

Question 5

D Environmental

A is incorrect because mesothelioma is not caused by a nutritional deficiency. **B** is incorrect because mesothelioma is not inherited from one's parents. **C** is incorrect because it is a physical disorder, not a psychological one.

Question 6

A

Nutritional	Environmental	Genetic
Rickets Beri beri	Rickets	Down syndrome Cystic fibrosis

Rickets is a nutritional disorder caused by a lack of vitamin D; rickets is also an environmental disorder caused by lack of sun exposure. Beri beri is a nutritional disorder caused by lack of vitamin B1. Down syndrome is a genetic condition caused by non-disjunction of chromosome pair 21 during meiosis. Cystic fibrosis is an autosomal recessive genetic condition. **B** is incorrect because Down syndrome is not an environmental condition. **C** is incorrect because cystic fibrosis is not an infectious disease and beri beri is not an environmental disease. **D** is incorrect because rickets is not genetic and Down syndrome is not infectious.

Question 7

B There will be an increase in the rate of death by non-infectious diseases.

The projections for the leading causes of death show increases in all categories of non-infectious diseases. **A** is incorrect because all infectious diseases show projected decreases in the rates of death. **C** is incorrect because there is an observable trend that rate of death due to infectious diseases will decrease and rate of death due to non-infectious diseases will increase. **D** is incorrect because there are projected changes to the rates of death. The data is shown as deaths per 100 000 rather than absolute values.

Question 8

C Non-infectious diseases

Non infectious diseases contributed the highest number of deaths. For example, ischaemic heart disease, stroke and chronic obstructive pulmonary disease (the top three causes of death) are all non-infectious diseases. **A** is incorrect because HIV/AIDS is the only disease listed that is caused by a virus, and this is currently the 8th highest (or 7th projected for 2030). **B** is incorrect because none of the diseases shown in this graph are nutritional. **D** is incorrect because few deaths on this graph are attributed to accidental death and are not close to being the highest cause of death (e.g. road injury is the 9th highest cause).

Question 9

C effect on phenotype.

The *ALDH* alleles cause differences to the phenotype (i.e. the physical expression). In this case, the *ALDH1* allele is normal and the *ALDH2* gene increases the risk of cancer. **A** is incorrect because both alleles are located in the same position in the genome. **B** is incorrect because both alleles would be at the same locus on the same chromosome in the nucleus of a gamete. **D** is incorrect because, while the sequence of bases in the alleles may differ and consequently change the sequence of amino acids, the amino acids are not part of the allele themselves.

Question 10

B Enzyme activity decreases if *ALDH2* is present.

If *ALDH2* is present, the graph shows that the amount of acetyl aldehyde is increased. The presence of acetyl aldehyde indicates decreased enzyme activity, preventing the acetyl aldehyde from being converted to acetate. **A** is incorrect because, although it is high in mice homozygous for the *ALDH2* gene, it is very low for those homozygous for the *ALDH1* gene. **C** is incorrect because enzyme activity decreases rather than increases (increased activity would show a reduction in the amount of acetyl aldehyde as it would have been converted to acetate). **D** is incorrect because enzyme activity would be highest in wild-type mice, indicated by low amounts of acetyl aldehyde.

Short-answer solutions

Question 11 (2 marks for each row completed)

Cause	Effect	Named disease with this cause
Genetic	Mutation of the *CFTR* gene on chromosome 7	Cystic fibrosis
Environmental	Inhalation of asbestos fibres that cause an inflammatory reaction in the lungs	Asbestosis

Other answers are possible.

Question 12

a $$\text{Incidence} = \frac{\text{Number of new cases}}{\text{Total population}} \times 100$$

$$= \frac{32}{10174} \times 100$$

$$= 3.15\%$$ (2 marks)

b $$\text{Prevalence} = \frac{\text{Previous cases of diabetes + new cases of diabetes − deaths of diabetics}}{\text{Population on 1 Jan 2021 + births − deaths}} \times 100$$

$$= \frac{472 + 32 - 21}{10174 + 120 - 58} \times 100$$

$$= \frac{483}{10236} \times 100$$

$$= 4.72\%$$ (3 marks)

Prevalence is the proportion of the population living with the disease. Marks would be given for:

- calculating the total number of people with diabetes, by working out the number of previous cases, adding new cases and removing diabetics who died during the year (1 mark)
- calculating the size of the population, by taking the known population from the previous year, adding births and removing all who died (1 mark)
- calculating the prevalence, by dividing the total cases by the total population size and multiplying by 100 to get a percentage (1 mark).

Question 13

a Mercury poisoning is an environmental disease, because it is caused by ingesting chemicals found in the environment (1 mark). Despite passing to women through their food, it is not a deficiency or excess of any nutrient and thus is not nutritional (1 mark). Although the mercury can pass from the mother to the child through the placenta, this is not an abnormality in the genes or chromosomes and so it is not a genetic disease.

b Infectious diseases are caused by pathogens (1 mark). Mercury poisoning is caused by ingestion of a chemical. Therefore, it is a non-infectious disease (1 mark).

Question 14

The graph indicates that the rate of melanoma increased for both males and females from 1972 to 1997. While the rates of melanoma were previously very similar for males and females, the rate was much higher for males by 1997 (1 mark), with approximately 52 cases per 100 000 population compared with 34 per 100 000 in females (1 mark). However, it is not possible to predict from the graph whether this trend would continue (1 mark), as the data is limited. To make a better prediction, other data should be included (1 mark), such as age of participants in the study, exposure to UV radiation and so on.

Question 15

a (4 marks)

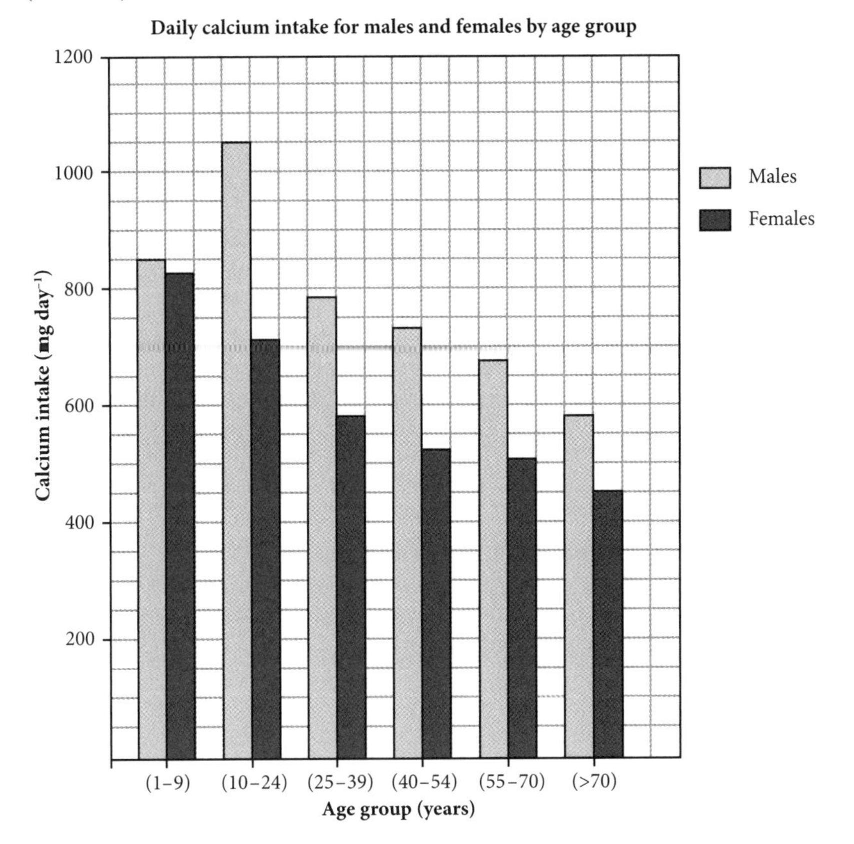

A column graph is appropriate, as the data is discrete rather than continuous. The graph should include:

- correctly labelled axes, including units
- scales on both axes (even on the *y*-axis and clearly labelling the age categories on the *x*-axis)
- a key to distinguish between males and females
- correctly plotted data
- an informative title that includes all variables (daily calcium intake, age and sex).

For 4 marks, all of the above must be included. If one element is missing, 3 marks would be awarded. If two elements are missing, 2 marks would be awarded. If more than two elements are missing, 1 mark may be awarded if there is some relevant information provided.

Note that if a line graph is provided, a maximum of 2 marks would be awarded if the title, axis labels and key are correctly completed.

b The graph shows that the calcium intake is higher for males than females in every age category (1 mark). The calcium intake decreases as age increases for females, and in males it initially increases to a peak in the 10–24 years age range, before declining after this point (1 mark). The average female only meets the required calcium intake level of 700 mg in the 1–9 years and 10–24 years categories, whereas males continue to receive sufficient calcium up to the 40–54 years age category. Over this age, calcium requirements are not met for males or females (1 mark).

SOLUTIONS – TEST 14

Test 15: Epidemiology

Multiple-choice solutions

Question 1

D Infectious and non-infectious disease, as well as accidents

Question 2

C may be inherited and affect the organism all their life.

A is incorrect because non-infectious diseases can develop in isolation. (If they were unable to do so, this would make quarantine an effective measure.) **B** is incorrect because non-infectious diseases are not caused by pathogens. **D** is incorrect because there are other techniques that can be used to treat and control non-infectious diseases.

Question 3

B An unpredictable variation in the data that leads to inconsistent results

A is incorrect because this describes sampling bias. **C** is incorrect because this describes prevalence/incidence bias. **D** is incorrect because this describes misclassification bias.

Question 4

D The incidence of diabetes is influenced more by age than physical activity.

A is incorrect because physical activity decreases the incidence of diabetes in all categories shown. **B** is incorrect because physical activity cannot cause diabetes. **C** is incorrect because there is no evidence to suggest that there is an increased rate of physical activity in older adults; there is simply a higher rate of diabetes regardless of physical activity status.

Question 5

A

Doll	Hill
Case-control	Cohort

All other options are incorrect because one or both studies are incorrectly classified.

Question 6

B The rate of colorectal cancer is continuing to rise in males but has begun to stabilise in females.

A is incorrect because the rate of colorectal cancer was previously similar in males and females (e.g. in 1985). **C** is incorrect because the total number of cases continues to increase. **D** is incorrect because the population does not have more males than females.

Question 7

A An intervention study

B is incorrect because a cohort study is used to study an exposed group and a control group to determine the cause of a disease. **C** is incorrect because a descriptive study is used to provide information about the pattern of a disease. **D** is incorrect because an analytical study is used to determine the likely cause of a disease.

Question 8

C The incidence and prevalence of melanoma is higher in more-developed regions of the world.

A is incorrect because the incidence and prevalence were slightly higher in males. **B** is incorrect because the incidence is lower in less developed regions. **D** is incorrect because people in Northern Africa and Polynesia can develop melanoma.

Question 9

B Pharmaceuticals

A is incorrect because non-infectious diseases are not caused by pathogens, so hygiene will have no effect. **C** is incorrect because non-infectious diseases are not transmissible, so there is no need for quarantine. **D** is incorrect because non-infectious diseases are not caused by pathogens, so vaccination will not be effective.

Question 10

C The rate of all cancers increased from 1982 to 2016.

A is incorrect because the rate is not decreasing in females. **B** is incorrect because the data shows an increasing rate and no specific case numbers. **D** is incorrect because there is no evidence to support this prediction.

Short-answer solutions

Question 11

The study conducted by this student is not well designed and the conclusion drawn is not valid (1 mark). The student had a very small sample, all attending the same gym, mostly female and all interested in being surveyed. This means her sample is unlikely to be representative of the population as a whole (1 mark). To improve this, she should collect data from a much larger sample (hundreds of people) from all backgrounds, different areas and so on, because if participants all live in the same area they are likely to be similar in other ways such as socioeconomic status and exposure to other factors (1 mark). Also, as all participants were in a gym, it is likely they exercise regularly (1 mark). To answer the question of whether exercise decreased the risk of CVD, she should also have surveyed a group of participants who did not exercise regularly (1 mark). Information on other risk factors such as smoking, diet and family history should also be gathered to identify any other possible causes.

This question can be approached in different ways. For full marks, a clear judgement about the methods used by the student should be included, along with at least two issues with the experimental design and at least two proposed changes that could be made.

Question 12

For example: Melanoma (a type of skin cancer) is most commonly treated by surgically removing the tumour and surrounding cells. This is followed by other strategies such as radiation or immunotherapy to ensure that all cancerous cells are removed and do not spread (1 mark). Advanced cases of melanoma that do not respond to these treatments may be managed through palliative care, pain management, counselling and complementary therapies to manage symptoms (1 mark). Future directions for the treatment and management of melanoma include further research into immunotherapy (to stimulate the immune system to fight the melanoma) and targeted therapies that use drugs to disrupt the pathways that cause the uncontrolled division of cells (1 mark). Vaccines are also being trialled whereby an antigen is injected to allow the immune system to more easily identify and deactivate the cancer cells (1 mark).

This question can be approached in many different ways, using any non-infectious disease as a named example. While no marks are allocated for simply naming the non-infectious disease, it is essential to provide this information to gain marks. Note that not all diseases have both treatment and management strategies; in this case, 2 marks could be awarded if treatment *or* management is discussed in sufficient detail. 2 marks are then allocated to discussion of future research, with two strategies for future research discussed.

Question 13 ©NESA 2021 MARKING GUIDELINES SII Q31

The study results show that when patients were in the 'no tablet' period they still displayed the symptoms of side effects (score 8). Taking placebo tablets doubled the severity of side effects (from 8 to 15.4) even though no active ingredients were present. Statin tablets and placebo tablets had very similar results (16.3 and 15.4). These data seem to indicate that side effects in patients taking statins may not be due to the active ingredients in statins but other factors.

However, these results cannot be regarded as valid, as only 60 patients were involved (millions of people take statins). Also, symptoms were monitored and recorded by the patients themselves in a qualitative way and cannot be regarded as reliable measurements.

The study should also have included a control group for comparison. The study and its results cannot be regarded as valid. However, the study is valuable as it indicates that it is worth following up with a large randomised control trial to confirm that statins do not cause side effects in most people. (6 marks)

This question can be approached in different ways. For full marks, the response must make clear reference to various aspects of the design of the study (e.g. sample size, control groups, scoring methods used), with an indication as to whether this will provide valid data (3 marks). Specific data should be included (1 mark), with correct interpretation of the data evident (1 mark). There should also be an evaluation of the validity of the study. This means that a clear judgement is made, supported by information from the stimulus (1 mark).

Question 14

a (5 marks)

1 A sample of 1000 people with colorectal cancer was selected and allocated to the test group (cases).

2 A second sample of 1000 people who did not have colorectal cancer was taken and allocated to the control group.

3 Participants in both groups were surveyed for information on medical background, exercise habits, diet and so on, and specifically on the amount of red meat and processed meat products they consumed.

4 This data was statistically analysed to determine whether there was a significant relationship between eating red meat and processed meat products, and developing colorectal cancer. The data was also studied for other possible causes of colorectal cancer.

Reponses to this question may vary. The sample answer uses a case-control study design, but there are other options (e.g. a descriptive study or cohort study). To gain full marks, information must be collected about the amount of red meat and processed meat consumed by participants (1 mark), a test group must be compared to a control group (1 mark), the sample size chosen must be suitably large to return reliable results (1 mark), the data must be statistically analysed to determine significance (1 mark), and the method used to collect data must be valid (e.g. use of valid surveys, data banks) (1 mark).

b Epidemiological studies can be used to determine the causes of infectious and non-infectious diseases as well as accidental deaths, in addition to who is affected, and where interventions may be most effective (1 mark). These studies allow public health authorities to direct funding and resources to where they are most needed and to evaluate current strategies (1 mark).

Question 15

The first graph shows that the amount of protein consumed per capita increased throughout the world from 1961 to 2011. However, the amount consumed per capita has increased only slightly in high-income ('developed') countries (from about 95 grams per day to about 103 grams per day), compared with a much larger rise in low-income ('developing') countries (from about 53 grams per day to about 75 grams per day).

The second graph indicates that foods containing protein (including some vegetables, meat, fish, eggs and dairy) have a much higher cost per unit of energy than foods high in carbohydrates (sugary processed foods and starches) as well as fats. For example, fruits and vegetables cost about $2.30 per 100 kJ and meat/eggs/fish about $1.55 compared with processed foods at $0.35, starches at $0.13 and fats at $0.05.

The increase in obesity may be attributed to a diet high in carbohydrates and low in protein, as the cost of healthier foods containing protein may be too high for many in low-income countries to afford, compared with much cheaper, high-energy carbohydrates. This would lead to much higher prevalence of non-infectious diseases (e.g. heart disease and diabetes) as well as infectious diseases with a lack of proteins to create antibodies to combat disease. (7 marks)

For this question, 4 marks would be allocated for the analysis of both graphs and identification of the trends in each, including specific data from each. 3 marks would be allocated to making a link between these trends and the rise in obesity in the developing world (e.g. the high cost of protein compared with the low cost of carbohydrates). Specific reference to data and/or categories in the graphs must be made.

Test 16: Prevention

Multiple-choice solutions

Question 1

A It would be quite effective because it suggests simple ways to avoid a known risk factor, encouraging the population to change their behaviour.

B is incorrect because this campaign does not include any legislation. **C** is incorrect because non-infectious diseases can also be prevented. **D** is incorrect because simple campaigns that are easily remembered and have clear instructions are more effective than detailed graphics.

Question 2

B Genetic testing of babies at birth for genetic conditions such as phenylketonuria or cystic fibrosis

Testing at birth is too late to prevent the occurrence of these non-infectious diseases. **A** is incorrect because array comparative genomic hybridisation is currently used to test embryos for genetic conditions. **C** is incorrect because genetic modification is currently used to add nutrients to food and prevent nutritional deficiencies. **D** is incorrect because genetic modification of viruses is currently used to produce vaccines.

Question 3

D The amount of vitamin A consumed will increase in all countries shown in the graph but will not allow people to reach the recommended daily intake in most of the countries.

A is incorrect because Japan already had the highest intake and the increase is much smaller than many other countries shown. **B** is incorrect because people in most countries will still not receive their required intake with Golden Rice. **C** is incorrect because while the daily requirements are not met, increasing the intake will have some benefit for individuals.

Question 4

C By the increase in the number of people aged 50 to 74 participating in the screening program

A is incorrect because the campaign only relates to bowel cancer. **B** is incorrect because the program does not target fibre intake. **D** is incorrect because asking for opinions on the effectiveness of a campaign is not a good indication of whether it has made an impact.

Question 5

C The government is responsible for passing legislation as well as funding research, regulating genetic engineering and developing public education campaigns.

A is incorrect because the government also has a role in funding and developing genetic engineering and public education. **B** is incorrect because the government also has a role in genetic engineering; it is not solely the responsibility of private organisations. **D** is incorrect because the government has an essential role in the prevention of non-infectious diseases as well as infectious diseases.

Short-answer solutions

Question 6

An effective educational program or campaign provides clear information to educate the public about the risk factors for a disease. The program should also contain strategies to avoid the risk factors in the hope that people will change their behaviour and avoid the risk factor. Successful campaigns contain a package of high-priority interventions and are evidence-based. One example of an effective educational campaign is the 'Slip, slop, slap, seek, slide' campaign, which explains the risk of developing melanoma and provides clear steps to avoid excessive sun exposure. Another example is the QUIT campaign, which explains the increased risk of lung cancer from smoking (e.g. through graphic images on advertisements and packaging) and interventions such as a national hotline to support those who are trying to quit smoking. (4 marks)

This question can be approached in different ways. Marks are given for:

- outlining at least two features of effective campaigns in detail (2 marks)
- providing at least two named examples, including a brief description of the strategies used (2 marks).

Question 7

Legislation involves governments passing laws to minimise the effects of risk factors in a population, such as banning a product or advertisement of the product, placing a tax on a risk factor. It can be seen that after each piece of legislation is introduced, the number of cigarettes smoked decreases (1 mark). For example, after the introduction of a ban on advertising in 1990, daily smoking prevalence decreased from 27.7% to about 23.7% in 1995 (1 mark). This trend has continued, with smoking being banned from certain venues, taxes introduced and further bans on advertisement of tobacco products resulting in a rate of 14.7% by 2015. However, the data provided is limited as it only provides the rates of smoking and does not include the incidence of lung cancer or deaths (1 mark). Despite this limitation, it is likely that the legislation introduced by the Australian Government will play a significant role in the reduction of lung cancer cases, as smoking is a known risk factor for lung cancer (1 mark).

This question can be approached in different ways. Responses should include: two correlations and/or limitations of the data provided; reference to specific data from the graph; and a clear judgement of the role of legislation in reducing lung cancer rates.

Question 8

CRISPRCas-9 technology can be used in people with sickle-cell anaemia to edit the genes that cause inactivation of foetal haemoglobin so it will continue to be made after birth. This prevents production of β-globin, which causes crescent-shaped cells in those with the condition (1 mark). The results from the trials with the two patients in the information provided suggest that it has been very effective, with 98% and 94% of red blood cells producing the foetal haemoglobin (1 mark) rather than the mutated β-globin. The patients also reported that they had not experienced any of the symptoms of sickle-cell anaemia, indicating that the treatment had been effective (1 mark). However, the data is limited, with the trials only being conducted in two patients (1 mark). A much larger sample size is required to be able to conclude that CRISPRCas-9 is suitable for treating sickle-cell anaemia. Despite the limitation of the data, it is clear that this technology shows potential to be a highly effective method of preventing sickle-cell anaemia (1 mark).

This question can be approached in different ways. The response should include an outline of how CRISPRCas-9 technology has been used to treat and prevent sickle-cell anaemia, at least two correlations and/or limitations from the stimulus provided, reference to specific data and a clear judgement of the effectiveness of the technology.

Test 17: Technologies

Multiple-choice solutions

Question 1

C It stimulates the auditory nerve.

A is incorrect because it does not amplify sound; this is the function of hearing aids. **B** is incorrect because it does not stimulate the eardrum, but rather bypasses it and stimulates the nerve directly. **D** is incorrect because it does not amplify vibrations in the cochlea – it bypasses it and connects to the nerve.

Question 2

B

1	*2*	*3*
Filtration	Reabsorption	Secretion

Question 3

A Filtration only

All other options are incorrect because the nephrons are unable to carry out the processes of reabsorption or secretion.

Question 4

D Auditory nerve

Question 5

A To transmit and amplify sound to the inner ear

The function of the tympanic membrane (labelled ii) is to transmit and amplify sound to the inner ear. **B** is incorrect because it is the vestibular canals that aid with balance. **C** is incorrect because it is the cochlea that converts mechanical waves to electrical energy. **D** is incorrect because it is the auditory nerve that sends electrical impulses to the brain.

Question 6

D the ability of the lens to change shape in order to focus light at varying distances.

All other options are incorrect because they describe processes other than accommodation.

Question 7

A It maintains the concentration gradient of the blood and the dialysis fluid, to speed up diffusion of waste.

B is incorrect because the same concentration of these substances is achieved by changing the dialysate fluid to prevent losing them through diffusion and osmosis. **C** is incorrect because it is the bubble trap that achieves this. **D** is incorrect because the counter-current flow does not control the temperature.

Question 8

C Cataracts

Cataracts are caused by clouding of the lens, so they can be treated by replacing the lens with an intraocular lens. **A** is incorrect because myopia is treated with spectacles/contacts or with LASIK surgery. **B** is incorrect because hyperopia is treated with spectacles/contacts or with LASIK surgery. **D** is incorrect because macular degeneration is caused by nerve damage and there is currently no effective treatment for it.

Question 9

C Males between the age of 20 and 60 are more likely to lose their hearing than females.

A is incorrect because there is no evidence that the cause of hearing loss is genetic. **B** is incorrect because males are likely to lose hearing after 70 if not prior to this. **D** is incorrect because there is no information provided that suggests the causes are environmental.

Question 10

C A person with a cochlear implant can hear most frequencies required for conversation, as long as the sound is loud enough.

A is incorrect because a person with a cochlear implant cannot hear any sounds quieter than 20 dB. **B** is incorrect because a person with profound hearing loss is better able to hear lower frequencies and only if quite loud. **D** is incorrect because a person with a cochlear implant can hear sounds at higher frequencies than with a hearing aid.

Short-answer solutions

Question 11

	Name of structure	Function
i	Lens	Refracts light to form a focused image on the retina. Changes curvature to focus light at varying distances. (1 mark)
ii	Iris muscles	Circular muscles that contract and relax to change the size of the pupil depending on light intensity (wider if intensity is low and narrower if high). (1 mark)
iii	Optic nerve	Sends electrical impulses from the retina to the brain. (1 mark)

Question 12 ©NESA 2021 MARKING GUIDELINES SII Q25

a (3 marks)

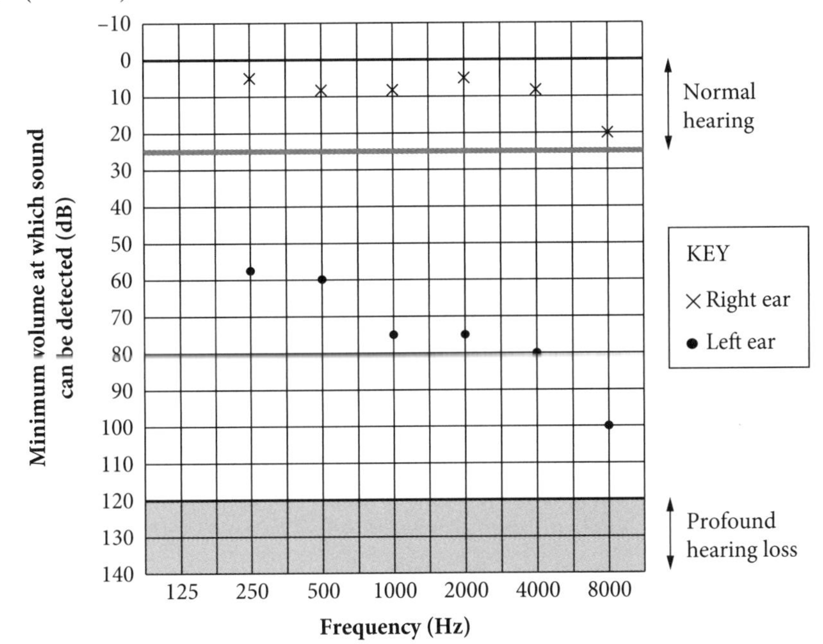

For this graph, 3 marks are given if the data is accurately plotted for both ears and an appropriate key is included. 2 marks are given if all data is plotted correctly but without an accurate key *or* if most data is plotted correctly and an appropriate key is included. 1 mark is given if some data is plotted.

b Right ear has normal hearing (1 mark). Left ear has a deficit/cannot hear at a normal level (1 mark).

c Bone conduction implants would be the most suitable technology for this patient. The blocked outer ear needs to be bypassed, and bone conduction works by detecting sound waves via a microphone and relaying it to a sound processor that converts the waves into vibrations. These vibrations are transferred directly to the cochlea in order for the person to hear the sound and so improve their hearing. (3 marks)

This question requires you to justify the choice of technology. You must provide reasons for the use of the named technology and why it is the most suitable choice. Your response should also outline how the technology works.

Question 13

a The kidneys are responsible for removing waste from the bloodstream (1 mark). A build-up of waste can increase the acidity of the blood, interfering with enzyme function and damaging cell structure (1 mark).

b For example: Nephropathy is a kidney disorder that is common in people with diabetes (1 mark). High levels of glucose in the blood put pressure on the nephrons as they need to filter more blood, leading to structural damage to the nephrons (1 mark). This allows large molecules that normally do not pass into the nephron (e.g. proteins and blood cells) to leak into the tubules and cause blockages (1 mark), which in turn cause further damage.

There are many examples that could be used to answer this question. 1 mark is given for naming/briefly describing the disorder, 1 mark for outlining the cause, and 1 mark for describing the effects.

c In renal dialysis, waste is filtered from the person's blood, preventing a build-up of waste in the blood that would increase acidity or damage cell structures in people whose kidneys are not functioning efficiently. However, dialysis machines can only perform the function of filtration; they cannot perform reabsorption or secretion. They are also not able to filter as much waste out as a functioning kidney can. Kidney transplantation occurs when a donated kidney is surgically transplanted into the body of a person whose own kidneys are not functioning efficiently. A successfully transplanted kidney will function normally, performing filtration, reabsorption and secretion at a constant rate. However, a donated organ must match the tissue type of a recipient precisely, and the immune system of the recipient must be suppressed to prevent rejection of the organ. Overall, renal dialysis is a life-saving technology that is able to sustain the lives of those without functioning kidneys, but it is limited in its ability to filter all waste and maintain the balance of substances in the blood. Kidney transplantation enables normal kidney function, but it can be difficult to obtain a suitable kidney and the process is not always successful. (4 marks)

This question can be approached in different ways. The response should show your understanding of what each technology involves (1 mark each) and the benefits and/or limitations of each technology (2 marks), and should provide a clear judgement of the effectiveness of each technology.

Question 14

a Myopia occurs when the focal length is too short and the image falls short of the retina. This can occur when the lens is too rounded or the eyeball is elongated (1 mark). As no image falls on the retina, no clear image is formed and sent via the optic nerve to the brain (1 mark).

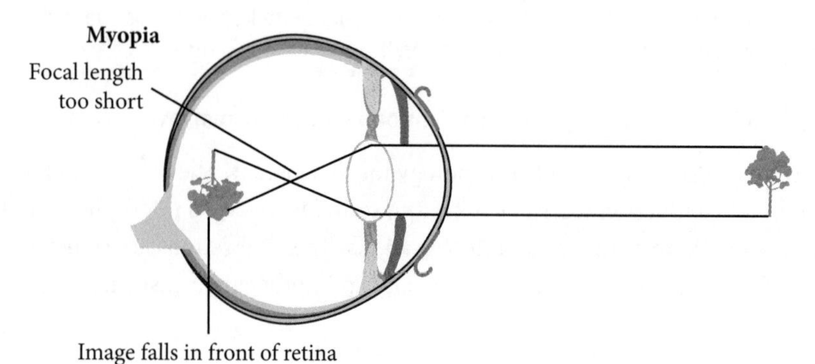

(1 mark)

There are other possible visual disorders that could be used to answer this question (e.g. hyperopia, cataracts).

b Myopia can be corrected with the use of concave lenses (1 mark). These bend the light rays outwards, diverging before they reach the eye (1 mark). This extends the focal length and causes the focused image to fall on the retina instead of in front of it (1 mark).

Correction of myopia

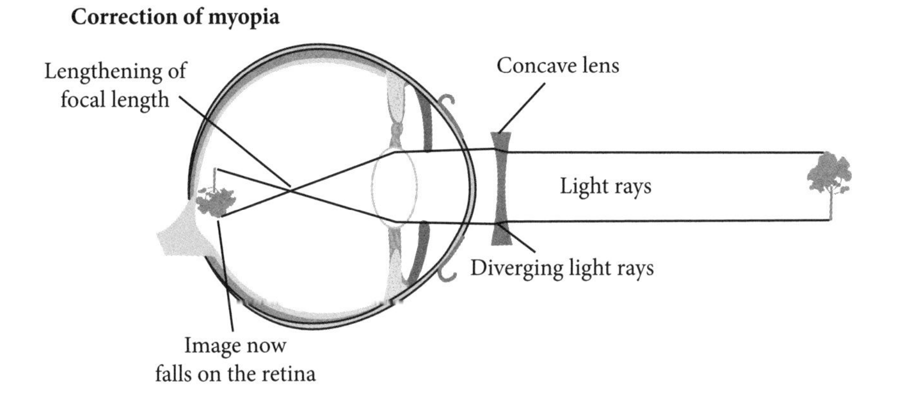

Note that the technology used must be appropriate for the visual disorder named (e.g. if hyperopia was named in part **a**, then convex lenses would be used). LASIK surgery is also an appropriate technology that could be used for either disorder. While a diagram is not essential for this response, it could be included to enhance your explanation.

Question 15

	Hearing aid	**Cochlear implant**
Location (1 mark)	Worn in the outer ears	An electrode is implanted in the cochlea. A microphone and speech processor are attached behind the pinna and transmit messages to the receiver.
Conditions in which it may be beneficial (1 mark)	Beneficial when there is some residual hearing as the sound is amplified	Beneficial when the person is profoundly deaf, but the cochlea is still functioning.
Benefits (1 mark)	Non-invasive (no surgery required) and relatively inexpensive	Allows people with profound hearing loss to hear at near normal levels. The implant is long-lasting
Limitations (1 mark)	Requires some residual hearing; will not benefit everyone with hearing loss. Wearers can experience feedback, have difficulty filtering out background noise or experience battery issues	Requires surgery to implant. Relatively expensive.

1 mark is allocated for each comparison (i.e. each row of the table). Although this question could be answered without the use of a table, it is often beneficial to construct a table for easy comparison, even when this is not explicitly asked for in the question.

SOLUTIONS – TEST 17

Practice HSC exam 1

Multiple-choice solutions

Question 1

C Fungi have a cell wall and protozoa do not.

Fungi have a cell wall, while protozoa, which are animal-like protists, do not. **A** is incorrect because fungi may be multicellular. **B** is incorrect because neither fungi nor protozoa have chloroplasts. **D** is incorrect because both fungi and protozoa are eukaryotic.

Question 2

C Reliability

Reliability is improved by increasing the sample size. **A** is incorrect because accuracy relates to measurements that are close to the true value, with little error in the measurements taken. **B** is incorrect because precision relates to the degree to which an instrument can be read repeatedly and reliably. **D** is incorrect because validity refers to experiments that test the hypothesis that was intended, with consistent results.

Question 3

D Vegetative propagation

Vegetative propagation is a type of asexual reproduction where new plants arise from the roots, stems, leaves or buds of adult plants. **A** is incorrect because tissue culture involves growing samples of tissue on an artificial culture medium in sterile conditions. **B** and **C** are incorrect because artificial pollination and sexual reproduction involve male and female gametes uniting to produce an offspring, which is not occurring in this scenario.

Question 4

D An mRNA strand

Transcription produces a strand of mRNA from DNA. **A** is incorrect because a polypeptide is produced as a result of translation. **B** is incorrect because while DNA is used as a template in transcription, it is not the product of the process. **C** is incorrect because amino acids have no role in transcription; they are involved in translation.

Question 5

B The leaf rolls up to increase humidity and reduce transpiration.

The leaf rolls up, which traps water vapour near the stomata; the increased humidity leads to decreased transpiration. **A** is incorrect because a waxy cuticle is waterproof to prevent evaporation; it is transparent, meaning it allows sunlight to pass through, rather than being reflected. **C** is incorrect because sunken stomata increase humidity and decrease transpiration. **D** is incorrect because leaf hairs are not involved in water absorption.

Question 6

A I

Follicle-stimulating hormone is produced by the pituitary. Levels rise at the start of the menstrual cycle, causing follicles within the ovary to mature. **B** is incorrect because II represents luteinising hormone levels, which rise dramatically before ovulation. **C** is incorrect because III represents oestrogen, and **D** is incorrect because IV represents progesterone.

Question 7

B a codon with the sequence CUU.

GAA is an anticodon on a tRNA. An mRNA with the complementary codon, CUU, would attach at point X. **A** is incorrect because A pairs with U in mRNA, not T. **C** and **D** are incorrect because the tRNA itself contains anticodons; the mRNA that attaches to tRNA contains codons.

Question 8

B X

Horse X is the donor of the nucleus, which contains the genetic information; therefore, Horse Z would be genetically identical to Horse X. **A** and **C** are incorrect because, while Horse W did donate an egg, its nucleus was removed, so it will not contribute any genetic information to Horse Z. **D** is incorrect because Horse Y is a surrogate only; it does not contribute any genetic information to Horse Z.

Question 9

D Australian intellectual property laws do not align well with Indigenous Australians' cultural view that no one can exclusively own the rights to a resource.

Australian intellectual property laws do not align well with the Indigenous Australian cultural view that no one can exclusively own the rights to a resource. **A** is incorrect because many bush medicines are as effective as those from other sources, as stated in the stimulus. **B** is incorrect because many Indigenous Australian people still use bush medicine. **C** is incorrect because Australian law does not currently require consultation with Aboriginal Elders to apply for the rights to use a resource.

Question 10

A a1/a1

Because cystic fibrosis is a recessive disorder, those with cystic fibrosis must have two recessive alleles. Of all the recessive alleles, a1 is by far the most frequent, so a1/a1 would be the most common genotype of cystic fibrosis patients. **B** is incorrect because the a2 allele has a very low frequency in the Australian population. **C** and **D** are incorrect, because the presence of a dominant allele, A, would mean individuals with these genotypes would not have cystic fibrosis.

Question 11

A The incidence rate of malaria was reduced due to the use of DDT.

The number of cases (incidence) of malaria started to rise in 1996 when DDT use was discontinued, peaking in 2000 before dramatically decreasing in 2001 when DDT was used again. **B** is incorrect because this graph does not give any data for prevalence; it is the incidence that was 64 000 in 2000. **C** is incorrect because the number of deaths (mortality) was much lower than the number of cases (incidence); note that there are two separate scales for incidence and mortality. **D** is incorrect, as despite some remaining cases and deaths, it is clear that the number was reduced when DDT was used.

Question 12

A Environmental

DDT exposure is an environmental disease, as people are exposed to the chemical in the environment. **B** is incorrect because the illness does not have a gene or chromosomal cause. **C** is incorrect because it is not caused by a pathogen. **D** is incorrect because it is not caused by a deficiency or excess of nutrients.

9780170465250

Question 13

What would you expect to observe in each of the three flasks?

C

Flask 1	Flask 2	Flask 3
No microbial growth	Microbial growth	Microbial growth

No microbial growth would occur in flask 1, as microbes in the air would be trapped in the S-bend of the flask's neck. Microbial growth would occur in flask 2, as microbes would be able to enter from the opening created when the neck was broken. Microbial growth would occur in flask 3, as the microbes trapped in the S-bend would mix with the broth and contaminate it. **A** is incorrect because no microbial growth would occur in flask 1. **B** is incorrect because microbial growth would occur in flask 3. **D** is incorrect because microbial growth would occur in flasks 2 and 3.

Question 14

C The chances of long-term survival are improved by receiving a transplant from either a living or a deceased donor.

Only about 10% of people receiving dialysis are still alive after 10 years, while those who receive a deceased or living donor transplant have a survival rate of 33% and 54% respectively. **A** is incorrect because dialysis does improve the chance of survival; without it, the person would have died within days of kidney failure. **B** is an incorrect interpretation of this information; effectiveness should be compared with the same time period. **D** is incorrect because a person with kidney failure cannot wait for a donation from a living donor; dialysis must be started immediately, and a donation from a deceased donor would increase the chance of survival and so should be received if available.

Question 15

B Bonds are formed between adjacent bases.

UV radiation has caused two thymine bases on the same strand to attach to one another, rather than to their complementary base on the opposite DNA strand. **A** is incorrect because thymine has not been duplicated; there are two thymine bases present both before and after UV exposure. **C** is incorrect because the sugar of one nucleotide forms a bond with the phosphate of the next nucleotide; this is not a result of mutation. **D** is incorrect because the two thymine bases shown are on the one DNA strand.

Question 16

B Sister chromatids separate during anaphase.

The photo shows sister chromatids being pulled apart during anaphase. **A** is incorrect because crossing over only occurs during meiosis, not mitosis. **C** is incorrect because two daughter cells are produced when the cytoplasm divides during cytokinesis, not telophase. **D** is incorrect because homologous chromosomes do not 'pair up' in mitosis, only in meiosis.

Question 17

D

	A	B
A	AA	AB
B	AB	BB

Two parents with the same phenotype can only produce offspring with three different phenotypes if the trait is inherited in a codominant or incomplete dominance pattern, and both parents are heterozygous. **A** is incorrect because there are four alleles shown, A, a, B, b, and the question states that only two alleles are involved. **B** is incorrect because the parents in this cross have different phenotypes, and the offspring of this cross would all have the same phenotype; this does not match the pedigree. **C** is incorrect because it appears to show a simple dominant-recessive relationship between the two alleles, which would result in only two possible phenotypes in the offspring.

Question 18

A Cancer cells carry unique antigens.

In order for an immune response to target cancer cells and not healthy body cells, the cancer cells must carry unique antigens. **B** is incorrect because self-antigens may be found on the surface of cancer cells. **C** is incorrect because there is no evidence that the patient's immune system is not functioning. **D** is incorrect, because a vaccine stimulates an immune response. The passage states that the vaccine is a useful treatment, so this immune response must be effective against the cancer cells.

Question 19

C cell division to produce more lymphocytes.

Vaccines stimulate the production of B and T lymphocytes by the process of cell division. **A** is incorrect because T cells do not produce antibodies. **B** is incorrect because helper T cells, not cytotoxic T cells, activate B cells. **D** is incorrect because B cells do not destroy melanoma cells directly; they do so via antibodies.

Question 20

A

Individuals with the genotype 2/3 are heterozygous, with one copy of allele 2, and one copy of allele 3. They would produce offspring in a 1 : 2 : 1 ratio, as shown in the following Punnett square:

	2	3
2	22	23
3	23	33

B is incorrect because it shows four offspring, all of whom are heterozygous. **C** is incorrect because it shows the right offspring but in the wrong ratio (this is possible but not the most likely result). **D** is incorrect because it shows that Child 4 has a copy of allele 1, which neither parent has.

Short-answer solutions

Question 21 ©NESA 2020 MARKING GUIDELINES SII Q23

a Point mutation/base substitution (1 mark)

b Chromosomal mutations (1 mark) involve changes to the number of chromosomes in the genome (1 mark).

Question 22

Type of mosaicism	Description	Impact on individual	Impact on individual's offspring
Gonadal	Mutation present in the germ-line cells of the gonads only	None	Offspring will inherit the mutation and have Marfan syndrome
Somatic	Mutation present in various organs but not the gonads	Individual will have Marfan syndrome	None
Mixed	Mutation present in various organs as well as the gonads	Individual will have Marfan syndrome	Offspring will inherit the mutation and have Marfan syndrome

To score 3/3, the table must be completely correct. To score 2/3, two rows (i.e. two mutation types) must be completely correct. To score 1/3, any one cell must be correct.

Question 23

a Pathogens that are food-borne or water-borne are often transmitted by the faecal–oral route. Pathogens will exit the body of the infected host in the faeces. Faecal matter may then be transferred to food via unwashed hands or flies (1 mark). If toilets are close to a water source, faecal matter may enter the water supply and be ingested by a new host (1 mark).

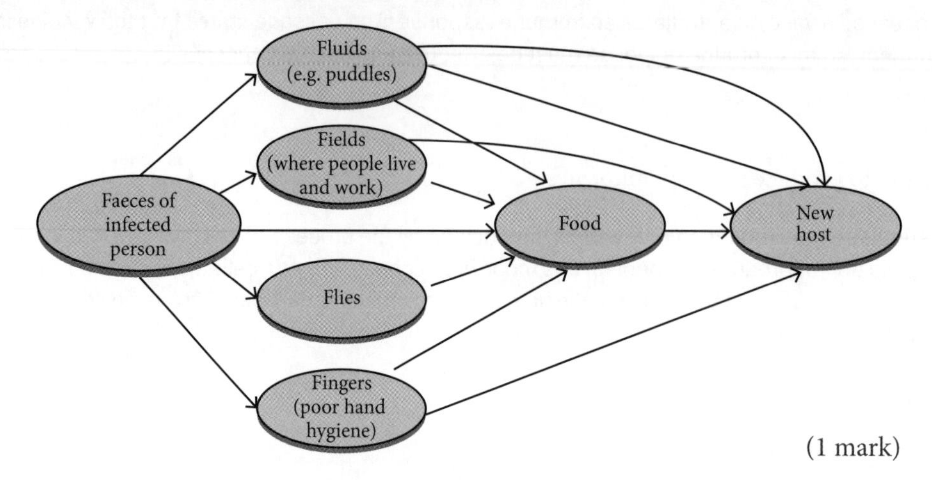

(1 mark)

b *Method:*

1 12 plates were prepared using a nutrient agar gel.

2 The workbench was sterilised using methylated spirit.

3 The bases of three plates were marked 'control', three were marked 'tap water', three were marked 'stream water' and three were marked 'rainwater tank water'. The date was written on the base of each plate.

4 Using a sterile swab, a sample of tap water was applied to the three 'tap water' plates.

5 Using a sterile swab, a sample of stream water was applied to the three 'stream water' plates.

6 Using a sterile swab, a sample of rainwater tank water was applied to the three 'rainwater tank water' plates.

7 All plates were sealed and placed in an incubator at 37°C for 48 hours.

8 After 48 hours, all plates were observed under a stereo microscope to observe and identify microbe growth on the plate.

9 Safety precautions: Plates were left sealed so that harmful microbes that may have been grown could not be inhaled or ingested. Hands were washed thoroughly at the conclusion of the experiment, to prevent accidental ingestion of microbes from water sources or from the plates. Plates were sterilised in an autoclave prior to disposal. (4 marks)

This question can be approached in different ways, as the method needs to include food *or* water sources. Alternative responses could use food sources such as yoghurt, cheese or bread. Some responses may even follow Koch's postulates, isolating a particular microorganism from a food source, growing it in pure culture, inoculating a healthy sample and then observing and identifying it as the original pathogen. To earn full marks, a response must include a logical method that tests for the presence of microbes in food or water, refer to sterile techniques, include a control as a basis for comparison and include repetition for reliability.

Question 24

Feature	Innate immune response	Adaptive immune response
Response time	Immediate response to an antigen	Response is slower (more than 96 hours)
Specificity	A generalised response to any antigen	A specific response to a particular pathogen
Memory	Has no memory	Has memory
Cells and structures	Physical and chemical barriers, macrophages, dendritic cells, neutrophils, natural killer cells, etc.	B cells and T cells

A table is not required, but is always a useful way to display a response if the question requires a comparison. To earn full (4) marks, at least two features must be compared, showing knowledge of both the innate and the adaptive response.

Question 25

a (4 marks)

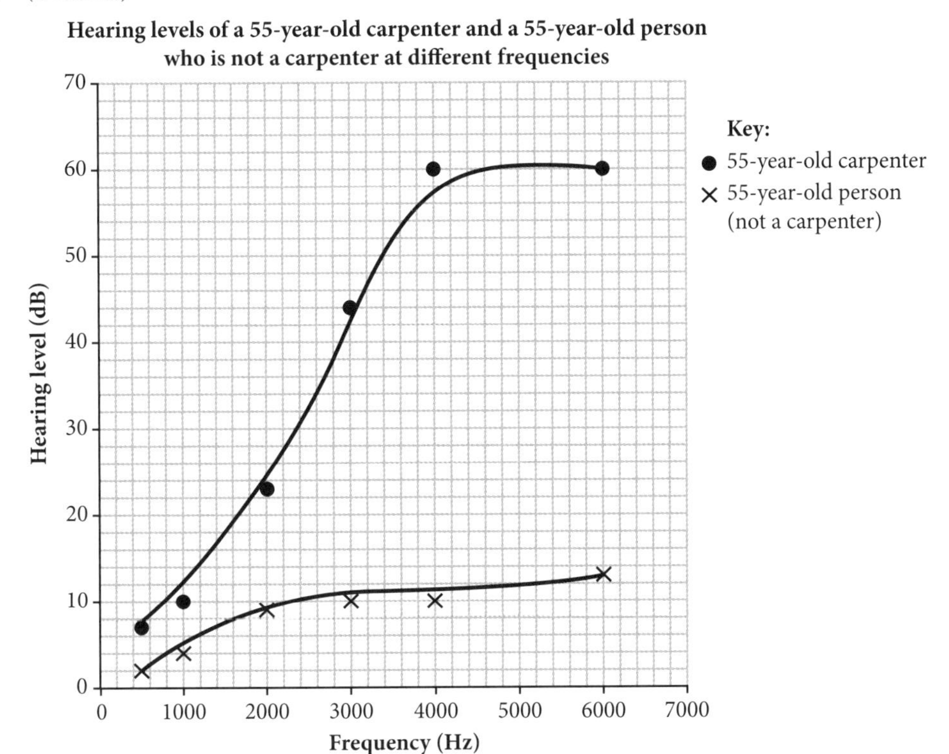

Answer must include:

- correctly labelled axes, including units (1 mark)
- an appropriate scale showing increments with equal value, unlike the column graph provided (1 mark)
- accurately plotted points, with the two sets of data clearly distinguishable (1 mark)
- accurately drawn curves of best fit (1 mark).

b The line graph is more useful in presenting the data (1 mark), because the data is continuous and the relationship between frequency and hearing levels of each person over time can be shown more accurately (1 mark).

c A hearing aid (1 mark) would be useful technology to use, as the carpenter would have some residual hearing. Hearing aids amplify the sound, making it louder (1 mark), so that the carpenter would be able to hear sounds of higher frequencies (1 mark). Hearing aids are less expensive and less invasive than other technologies.

This question can be approached in different ways. For full marks, the response must include a benefit of the named technology and how it would enable hearing to be improved. While hearing aids are the most likely technology to be used in this situation, answers could include cochlear implant or bone conduction if the answer includes a justification for their use.

Question 26

Mutations are changes in genetic information. Mutations would have generated new alleles and increased the size of the koala gene pool on Kangaroo Island. If a new allele provided a survival advantage, it would have become more common over time; if it was a disadvantage, it would have disappeared from the population. (2 marks)

Gene flow is the transfer of alleles from one population to another, usually through migration. Gene flow generally increases the amount of variation within the gene pool of a population. If the Kangaroo Island koalas were genetically isolated from those on the mainland, gene flow would have had no impact on the gene pool; however, if humans intervened between 1920 and 2020 by transferring koalas from the mainland to the island, then gene flow would have had a positive impact, by increasing the size of the gene pool. (2 marks)

Genetic drift occurs when chance events lead to random changes in the frequency of alleles within the gene pool. Genetic drift has the potential to decrease the size of the gene pool, particularly when the population is small. The bushfires of 2019–2020 had a negative impact on the gene pool, because they drastically reduced the Kangaroo Island koala population, thus reducing the size of the gene pool and leaving the koala population vulnerable. (2 marks)

2 marks are allocated to each process: mutation, gene flow and genetic drift. For each process, 1 mark is awarded for providing a sound understanding of the process itself; this does not need to be in the form of a definition, as in the sample answer; however, the understanding must be clearly evident in your response. The second mark is for assessing the impact that the process had on the gene pool of the Kangaroo Island koala population.

Question 27 ©NESA 2012 SAMPLE ANSWERS SII Q26

The design of this study cannot validly lead to a link between disease and its likely causes. A valid questionnaire is good but the number of subjects is low and only confined to the workplace. The sample should be larger and broader (1 mark). Ideally, the study should have a variety of equal categories, e.g. age, ethnicity, not just equal male : female (1 mark). Participants should not be eliminated on the basis of their answers, as this reduces the scientific validity (1 mark). Any checks should be consistent, with a definite purpose related to the study, e.g. lungs checked (1 mark). The final data should be peer reviewed for publication (1 mark).

To earn full marks, the response must analyse the methodology by identifying strengths and/or weaknesses in each step of the study from the information provided. The response should show a good understanding of the features of a sound epidemiological study.

Question 28

a Bubonic plague was primarily transmitted by vectors in the form of fleas carried by rodents, enabling transmission of the disease from one host to another (1 mark). This would increase the chance of transmission and also provide an animal reservoir that allowed the disease to survive outside the human host (1 mark). Mobility of humans travelling by ship throughout Asia and Europe moved the disease from one location to another (1 mark). The disease spread much further than it would have if it had simply gone from host to host, as infected humans travelled from port to port, and city to city, with rodents aboard the ships also travelling and spreading the disease to a new area. The map shows that it originated in Asia in 1346 and had spread throughout most of Europe by 1351 (1 mark).

This question can be approached in different ways. The 4 marks would be evenly divided between the role of mobility (2 marks) and mode of transmission (2 marks). Answers should not merely identify the mode of transmission, but provide some detail regarding how that increases transmissibility. Reference should be made to the stimulus when discussing the role of mobility.

b Plague doctors wore leather clothing, as well as a beak-shaped mask. The leather clothing would have prevented some contact with the pathogen. The beak-shaped mask was filled with herbs as there was a belief that the bad smell of plague caused infection. The mask would have reduced the chance of inhaling the pathogen. However, the equipment worn would not have been as effective as modern-day personal protective equipment (PPE). Modern-day equipment is usually disposable, so there is no risk of transmission later, and surgical and N95 masks have much smaller weave, which prevents entry of bacteria and are well-fitted to the face.

Medical inspection by a plague doctor would allow confirmation of suspected cases, who were then required to isolate in homes or hospitals. This would prevent the infected person from transmitting the disease to a healthy person as they were not in close contact with them. Isolation for 40 days would allow enough time to pass so that the person was no longer infectious before they were allowed out of isolation. This would have been quite effective; however, it would not prevent infection of others in the home or of people tending to the sick in plague hospitals.

With a modern-day understanding, they could have conducted proper diagnostic tests to confirm infection and then treatment with antibiotics to cure the disease. Properly fitted medical-grade PPE would have prevented transmission to those who were nursing sick patients. An understanding of the role of rodents and fleas as vectors could also have helped to prevent transmission, including using pesticides to remove vectors from the environment. (5 marks)

This question can be approached in different ways. For full marks, the response must include a discussion of *two* of the measures used, with a judgement of their effectiveness, and identification of at least two other measures that could have been implemented if people had had a modern-day understanding of disease control.

Question 29 ©NESA 2014 MARKING GUIDELINES SII Q29

The offspring would not be a clone of the female, because the offspring and the female would not be genetically identical.

The offspring was produced using two ova. The nuclei of the two ova are different from each other and from the nuclei in the female's body cells. The ova were produced by meiosis of a body cell of the female.

During meiosis, the processes of independent assortment of chromosomes and crossing over occur. In crossing over, segments of DNA swap positions on homologous chromosomes so new gene combinations are possible. Independent assortment results in different sorting of chromatids into ova. This results in genetically unique ova.

If these two genetically unique ova fuse, there will be a new unique combination of genetic information in the offspring that is not identical to that of the female who produced the ova (eggs). (5 marks)

There are a number of ways to structure this response. For full marks, the response must correctly assess that the offspring will not be a clone of the female, because a clone is genetically identical. The response must explain why there would be genetic difference between parent and offspring, in terms of the role of meiosis (including crossing over and independent assortment), as well as the combining of gametes (via fertilisation). The response must also link this explanation to information provided in the stimulus. If an incorrect assessment is made (i.e. that the offspring is a clone), the maximum mark is 2, provided two correct statements about cloning or sexual reproduction are included in the response.

Question 30

An understanding of disease, disease transmission and the immune response has allowed scientists to develop measures and technologies that prevent, treat and control the spread of disease. Control of disease has included development of personal protective equipment, limiting mobility, implementing isolation guidelines, implementing hygiene guidelines and use of antimicrobial substances in the environment (e.g. disinfecting surfaces). These measures required an understanding of disease transmission; for example, knowledge of airborne diseases or those carried in droplets has led to use of masks and other PPE, while knowledge that pathogens can survive outside the host on hard surfaces has led to development of hand hygiene and cleaning guidelines.

Vaccinations have been developed that prevent a person from getting a disease, or decrease the severity of the symptoms, which decreases the likelihood of transmission. Vaccinations work by stimulating the adaptive immune response, stimulating proliferation of B cells, which differentiate to form plasma cells that secrete antibodies, and B memory cells, which allow a rapid response to the pathogen upon secondary infection. An understanding of the adaptive immune response was required in order to safely develop vaccines that stimulate a response without causing disease. An understanding of the pathogens that cause disease has enabled scientists to develop attenuated strains or inactivated viruses that are not harmful to the host.

Treatment of diseases has improved, including the development of antibiotics to treat bacterial infections and antivirals to treat viral infections. An understanding of the pathogen is required in order to destroy it or prevent reproduction/replication. Understanding how pathogens reproduce in cells is also necessary, as this allows scientists to develop chemicals that prevent them from leaving a host cell.

While vaccination and antibiotics have saved countless lives, they can both lose effectiveness over time. Bacteria may develop resistance to antibiotics. This occurs because some bacteria in the population will be naturally resistant to the substance and if the full dose of antibiotics is not administered, a resistant population will develop and the antibiotics will no longer be effective against those bacteria. The overuse and misuse of antibiotics as well as the excessive use of antimicrobial substances in the environment have created resistant strains of bacteria, requiring scientists to continually find new drugs that can destroy bacteria while causing minimal harm to the host.

Vaccines result in the production of antibodies and B memory cells. Over time, the levels of these drop dramatically, so many vaccines require two or more doses, as well as boosters at a later time. Additionally, many pathogens mutate rapidly, which can mean that a vaccine is no longer effective against a pathogen as the memory cells will not recognise the pathogen and the antibodies will not be able to deactivate it.
(9 marks)

This question can be approached in different ways. The response should include:

- discussion of measures or developments for prevention, treatment and control
- a link between the measures and scientific understanding that allowed for their development
- thorough knowledge of the immune response and how the control measures work
- an explanation of antibiotic resistance, including reasons for its development
- *one* explanation for reduced effectiveness of vaccines – i.e. antibody levels drop, so second doses or boosters are required, *or* virus mutates, so the vaccine no longer produces memory cells that can recognise that pathogen.

Five marks would be allocated to the discussion of prevention, treatment and control, linked to the scientific understanding required for development. Four marks would be allocated to the explanation of antibiotic resistance and reduced effectiveness of vaccines.

Question 31

a (3 marks)

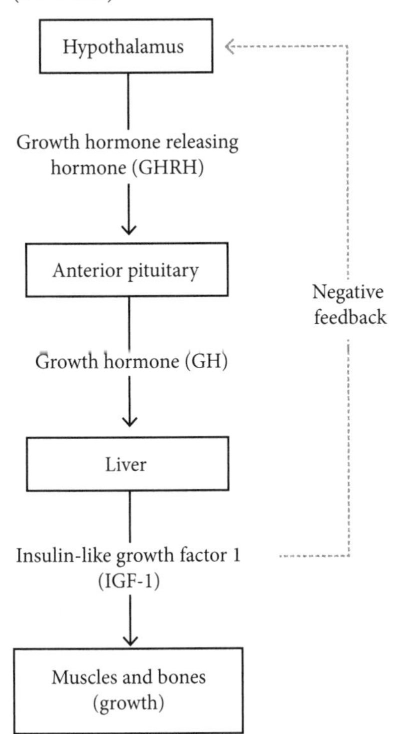

b **i** The woman must have type II congenital GHD, which is autosomal dominant. (1 mark)

We can assume, because the male has no family history of the disease, that he does not carry any affected alleles.

It cannot be type I (autosomal recessive) because all offspring would inherit a normal, dominant allele from their father. Therefore, it would be impossible to have an affected daughter:

	R	R
r	Rr	Rr
r	Rr	Rr

– i.e. no affected children (1 mark)

It cannot be type III (X-linked recessive). All sons would inherit their one X chromosome from their mother and would be affected, but because all females would inherit a normal, dominant allele from their father, it would be impossible to have an affected daughter:

	X^R	Y
X^r	X^RX^r	X^rY
X^r	X^RX^r	X^rY

– i.e. only sons are affected (1 mark)

It must therefore be type II (autosomal dominant). The mother, who is affected, could be either homozygous (RR) or heterozygous (Rr). In both instances, it is possible to have an affected daughter:

	r	r
R	Rr	Rr
R	Rr	Rr

– i.e. all children are affected (1 mark)

	r	r
R	Rr	Rr
r	rr	rr

– i.e. 50% of children affected, both males and females possible (1 mark)

For full marks, the response must correctly identify type II congenital GHD. The conclusion must be justified by proving that it is *only* possible to have an affected daughter if the woman has type II GHD. The justification must include Punnett squares showing the outcomes of each cross to support the response. Because this question is worth 5 marks, it is not enough to prove that type II 'works', ignoring the others. A response that does this would achieve a maximum of 3 marks (1 mark for stating the correct type, 1 mark for an explanation of why this type is possible, and 1 mark for a correct Punnett square to support the explanation).

ii Because a child with type IA does not produce any growth hormone (GH), the body's immune system would recognise GH as foreign (1 mark). The treatment would initially be effective, as the GH would remain in the body for long enough to act on target organs (1 mark). With repeated GH injections, the level of anti-GH antibodies increases, and GH would be destroyed before it can take effect (1 mark). The treatment is no longer effective; hence, type IA children remain short into adulthood.

Those with type 1B do produce some of their own GH. This means that the immune system would not recognise GH as non-self, and would not produce destructive antibodies against it (1 mark). It could remain in the body, where it acts to stimulate growth of bones and muscles, leading to normal height in adulthood.

c A gene is a section of DNA that codes for a polypeptide. Mutations are changes in the structure or sequence of DNA, which can change the resulting polypeptide and cause genetic diseases and disorders. These mutations, and the genetic diseases and disorders that they cause, can be passed from parent to offspring.

Treatment

Recombinant DNA technology involves inserting a gene from one species, such as the gene for human GH, into the genome of another, such as *E. coli* bacteria. In order to do this, scientists must understand which particular gene (e.g. the *GH1* gene) codes for which polypeptide (GH). They must be able to isolate a working copy of the desired gene from a human cell or build a synthetic copy of the gene in the laboratory, and splice this copy into a bacterial plasmid. To do this, they must understand the physical structure of human and bacterial DNA, and how the enzymes required for this process operate (such as restriction enzymes and DNA ligase).

With this understanding of gene structure and function, scientists are able to create transgenic bacteria that can produce human hormones such as protropin, insulin and factor VIII. These bacteria can be used to provide cheap, accessible and plentiful hormone replacements that can treat diseases such as GHD, diabetes and haemophilia. An understanding of how these genetic conditions are inherited (e.g. autosomal, recessive, sex-linked) is also important, as this allows for a child with hormone deficiency to be promptly diagnosed and treated as quickly as possible.

Prevention

IVF involves the retrieval of ova from a female and sperm from a male, so that ova can be fertilised outside the body (in vitro). If a potential parent has or carries a genetic disease, such as Huntington's disease or cystic fibrosis, embryos produced by IVF can be screened for the disease prior to implantation, by pre-implantation genetic diagnosis. Only embryos that are free of the condition are implanted; in this way, the disease is prevented.

Scientists must understand the genetic cause of the disease in question, e.g. mutations in the *GH1* gene cause GHD, or mutations in the *CFTR* gene cause cystic fibrosis (CF). They must also understand the inheritance pattern of the disease (e.g. autosomal recessive), so that they can determine which embryos will have the disease and which will not. Without this understanding of genes and their inheritance, this technique could not be used to successfully prevent non-infectious diseases such as CF. (8 marks)

This question has many components, so it would be beneficial to break the question down and highlight key features of the question and their implications before writing a response. For full marks, a response must address all the following components.

- The verb is *explain*. The response must link cause and effect.
- The response must explain 'how our understanding of genes *and* inheritance has led to genetic technologies' (note the plural – i.e. at least two technologies).
- There are many examples of technologies that would be suitable, e.g. gene therapy, CRISPR gene editing, therapeutic cloning, artificial insemination, gene cloning/PCR.
- The response must link genetic technologies to 'improved treatment *and* prevention of non-infectious diseases' (i.e. you must address both treatment and prevention, and include at least two examples of non-infectious diseases).
- Non-infectious disease is the limiting factor in this question – the response should not discuss infectious disease.
- The response should incorporate information from the stimulus.

This question can be approached in many ways. Subheadings (as shown in the sample answer) or even a table would help to ensure a logical and coherent response, as it can be easy to go off track in this question, because it is quite broad.

Practice HSC exam 2

Multiple-choice solutions

Question 1

D It increases the likelihood of the successful union of gametes.

Internal fertilisation deposits sperm within the reproductive tract of the female, which increases the chances of fertilisation. **A** is incorrect because it does require sexual intercourse. **B** is incorrect because some aquatic organisms (e.g. whales and dolphins) reproduce by internal fertilisation. **C** is incorrect because the offspring will be genetically unique, because they inherit half their genetic information from each parent.

Question 2

A

Label	Structure	Function
i	Cornea	Refraction of incoming light rays

The cornea is responsible for approximately two-thirds of all refraction of light in the eye. **B** is incorrect, as the lens changes shape to focus light at varying distances but does not control how much light enters the eye. **C** is incorrect because label iii indicates the sclera. **D** is incorrect because label iv indicates the vitreous humour.

Question 3

C Authorities should inspect wheat seeds as they enter ports, keeping them quarantined for the entire incubation period.

Inspecting seeds as they enter ports and keeping them quarantined for the entire incubation period should allow signs of infestation to appear, preventing the release of diseased seeds into Australia. **A** is incorrect because randomly inspecting crops that are already in Australia would occur too late to prevent the disease entering Australia. **B** is incorrect because authorities cannot trust that all farmers and importers would answer truthfully, or disease may develop after seeds have been shipped. **D** is incorrect because flour and baked goods do not pose a risk; it is the seeds or plants that will carry disease that could spread to Australian crops.

Question 4

D Transgenic

A transgenic organism is one that has had a gene or genes from another species inserted into its genome. **A** is incorrect because a clone is an organism that is genetically identical to another. **B** is incorrect because a hybrid is an organism that results from breeding individuals from genetically distinct populations. **C** is incorrect because plants are eukaryotic, not prokaryotic; both prokaryotic and eukaryotic organisms can have foreign genes inserted into their genome.

Question 5

B Antigen

Antigens are molecules that trigger an immune response. **A** and **C** are incorrect because antibodies (and immunoglobulins, which are a class of antibodies) are produced by the immune system in response to foreign substances. **D** is incorrect, because a pathogen is a disease-causing agent, such as a bacterium, virus or fungus. Cry9C is a protein, not a disease-causing organism or a microbiological agent.

Question 6

B Gene flow

Gene flow occurs when new individuals enter a population, or existing individuals leave a population, resulting in a change in allele frequency. It is likely that cross-pollination occurred between closely planted StarLink and unmodified crops, introducing the *Cry9C* gene into the unmodified population. **A** is incorrect because artificial pollination requires humans to transfer pollen from one plant to another. Cross-pollination between the StarLink and the unmodified crops occurred naturally (via wind or animals), rather than via human intervention. **C** is incorrect because genetic drift occurs when events such as natural disasters cause a change in the frequency of alleles that already exist within a population; it does not introduce new alleles into a population. **D** is incorrect because mutation cannot generate an entirely new *Cry9C* gene (especially one from another species).

Question 7

C The lens does not refract incoming light rays enough, so the image falls behind the retina.

A is incorrect because the lens does not refract incoming rays too much, but rather not enough. **B** is incorrect because the lens does not refract incoming rays enough and the image does not fall in front of the retina, but rather behind it. **D** is incorrect because the image falls behind the retina, not in front of it.

Question 8

C Resistance to drying out and changes in oxygen concentration

Resistance to drying out and changes in oxygen concentration will prevent the virus from being destroyed by the change in environmental conditions outside the host. **A** is incorrect because a low tolerance to antimicrobial substances in the environment would decrease the pathogen's chance of survival. **B** is incorrect because being heavy would make the particles fall rather than remain suspended in the air. **D** is incorrect because surface proteins on the virus may help to adhere to the host cell, but do not increase survival outside the host.

Question 9

A Pathogens are destroyed by antimicrobial chemicals on the skin.

The skin produces antimicrobial substances that destroy pathogens. **B** is incorrect because viruses are able to tolerate body temperature, and certainly the lower temperature of the skin. **C** is incorrect because viruses on a person's hands or other skin surfaces may be ingested or inhaled after they adhere to the skin. **D** is incorrect because handwashing may remove pathogens from the skin, but will not prevent adherence afterwards.

Question 10

C DNA replication

The process shown is polymerase chain reaction, where a single copy of a DNA strand is amplified to produce many copies. This process is similar to that of DNA replication, which occurs within the cell nucleus. **A** is incorrect because mitosis does not copy the genetic information (as shown in the graph); rather, it divides the copied information into two. **B** is incorrect because fertilisation involves the union of sperm and ovum, which is not happening here. **D** is incorrect because polypeptide synthesis involves transcription and translation, neither of which is shown here.

Question 11

B Denaturation

In PCR, denaturation occurs when DNA is heated to very high temperatures. This breaks the hydrogen bonds between nitrogenous bases and separates the double-stranded DNA into two single strands. **A** is incorrect because annealing occurs when the temperature is lowered and DNA primers attach to the DNA. **C** is incorrect because division is not a process that occurs in PCR. **D** is incorrect because extension occurs at an intermediate temperature, where polymerase enzyme 'builds' two new DNA strands.

Question 12

B

Indirect contact	Direct contact	Vector
Glandular fever Hepatitis Influenza	Glandular fever Ringworm Zika virus	Malaria Zika virus

Ringworm and Zika virus require physical contact between individuals. Glandular fever is transmitted by indirect contact or direct contact. Hepatitis and influenza do not require physical contact; instead, they result from contact with a contaminated material, surface or object. Malaria and Zika virus are transmitted by an intermediate organism, called a vector. **A** is incorrect because ringworm is not transmitted indirectly and hepatitis is not transmitted directly. **C** and **D** are incorrect because Zika virus is not transmitted by indirect contact, hepatitis is not transmitted by direct contact and ringworm is not transmitted by vector.

Question 13

A It is both a genetic disease and an infectious disease.

It is genetic because it is encoded by a gene, but also infectious as misfolded protein (prion) can be transmitted from one host to another by ingestion. **B** is incorrect because it is not only genetic, but also infectious. **C** is incorrect because it is infectious, with the prion being the pathogen, despite being non-cellular. **D** is incorrect because it is the misfolded protein that is the pathogen, not the normal protein.

Question 14

D Tortoiseshell females and orange males

If tortoiseshell cats have both orange and black fur, the two alleles must be codominant. An orange female would be X^OX^O and a black male would be X^BY. All female offspring would be X^OX^B (tortoiseshell), and all males would be X^OY (orange). **A** is incorrect because no females will be orange, as they will all inherit an X^B from the father, making them tortoiseshell. **B** is incorrect because males only have one allele for fur colour, so cannot be tortoiseshell, as this requires one orange and one black allele. **C** is incorrect because orange females need to have two X^O chromosomes, one of which comes from the father, but the father cat in this case can only provide an X^B chromosome.

Question 15

D The student is incorrect. Cytokines are injected into the infected cells by cytotoxic (killer) T cells to destroy the infected cell.

The student is incorrect, as a pathogen can be destroyed once inside an infected cell, if cytotoxic T cells inject cytokines into the infected cell. **A** is incorrect because the student's statement is not correct; pathogens can be destroyed even after entering a cell. **B** is incorrect because antibodies are not injected into cells; they work in the bloodstream or extracellular fluid. **C** is incorrect because cytotoxic T cells can destroy pathogens inside cells, not only outside the cells.

Question 16

D Vitamin A deficiency is the most prevalent micronutrient deficiency in the children studied.

Vitamin A deficiency has the highest overall average prevalence for all age groups. **A** is incorrect because, while no data is shown in the graph, it is unlikely that it could affect children younger or older than this but not children 18–23 months old. **B** is incorrect because, although no data is shown, children under 6 months of age can still have some deficiencies. **C** is incorrect because the prevalence does not increase with age; in fact, the highest prevalence for each deficiency is in the youngest age group.

Question 17

B Sperm may contain 11 or 12 chromosomes.

Males only have one sex chromosome. As a result of meiosis, only 50% of male gametes will contain the sex chromosomes (and will have 12 chromosomes in total). The other half will have no sex chromosome (and therefore will only have 11 chromosomes). **A** is incorrect because male offspring are still produced by the union of a male and female gamete. **C** is incorrect because both types of male gametes (those with and without a sex chromosome) can fertilise an egg; if a sperm without a sex chromosome fertilises an egg, the offspring will be male; if a sperm with a sex chromosome fertilises an egg, the offspring will be female. **D** is incorrect because males have a copy of the sex chromosome, so they will inherit traits carried on the sex chromosome.

Question 18

A It would reduce reliance on traditional pesticides.

The GM fungus would kill its host more quickly; this would mean less need for pesticides to control the locusts. **B** is incorrect because if the fungus is more expensive, this would be a disadvantage of its use, not a potential benefit. **C** is incorrect because a loss of biodiversity would also be a disadvantage. **D** is incorrect because requiring specific environmental conditions is a disadvantage, as it would limit the potential use of the fungus to areas where these conditions are present.

Question 19

B The incidence rate was 7.87 per 100 000.

The number of new cases was 980 and the susceptible population was 12 450 000 (49.8% of the total population of 25 million, as only men are susceptible). $\frac{980}{12\,450\,000} \times 100\,000 = 7.87$. **A** is incorrect because this number would be obtained by using a population size of 25 million, not recognising that only males will be susceptible to this cancer. **C** and **D** are incorrect because prevalence rate cannot be calculated from the information provided; prevalence is all new cases plus pre-existing cases of the disease.

Question 20

D There is a 50% chance that their offspring will have a copy of chromosome 21 attached to chromosome 14.

The diagram shows the possible gametes the carrier male can produce, and the possible offspring when these gametes fertilise the gametes of the normal female.

Because the carrier male has a balanced translocation (one normal chromosome 14 and one chromosome 14 with chromosome 21 attached, as shown in the diagram), half of all possible gametes he produces will contain a chromosome 14 with chromosome 21 attached. If 50% of the male's gametes contain the affected chromosome 14, then there is a 50% chance that the offspring will possess the affected chromosome. **A** is incorrect because there is only a 25% of offspring being carriers. **B** is incorrect because there is a 25% chance of the offspring having translocation Down syndrome. **C** is incorrect because there is only a 25% chance of the offspring being monosomic (having only 45 chromosomes).

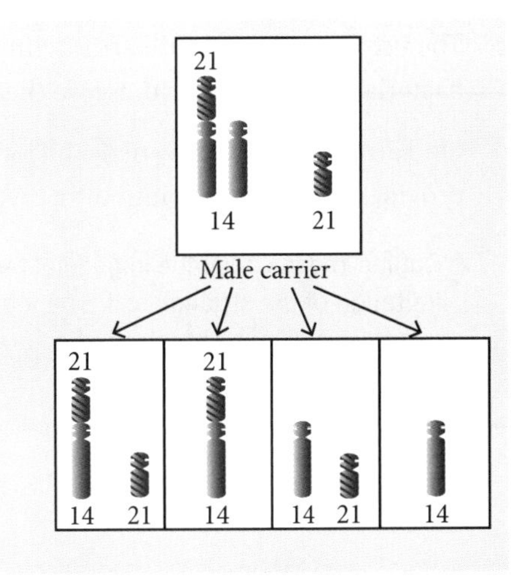

Short-answer solutions

Question 21 ©NESA 2021 MARKING GUIDELINES SII Q21

a (2 marks)

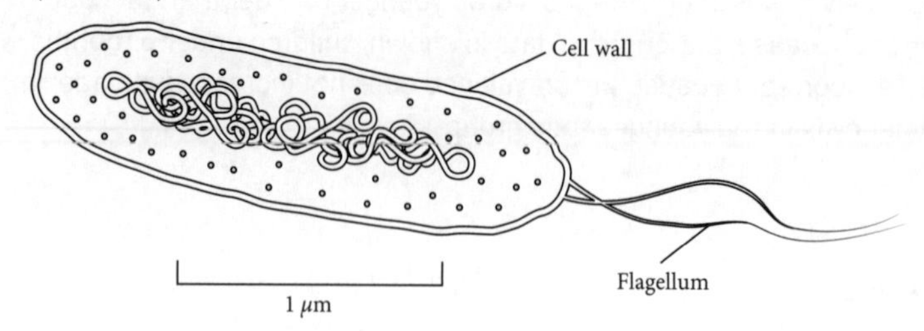

1 mark is allocated for each correct feature. Other features include plasma membrane, absence of membrane-bound organelles, DNA.

b

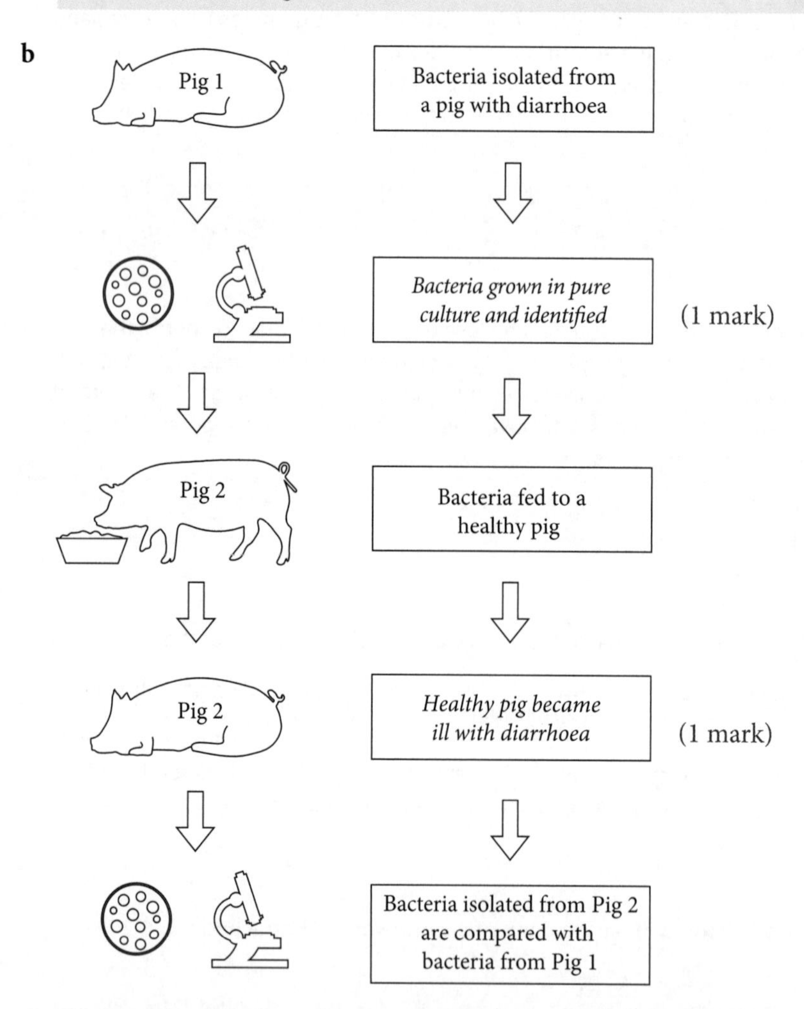

c The use of antibiotics on Farm 1 eliminates the disease quickly but may induce resistance in the bacteria with longer-term use, so the strategy will become less effective.

On Farm 2, improved hygiene, including removal of rats and mice, is slow to eliminate the disease but provides a long-term solution that will prevent future outbreaks. (3 marks)

'Outline' means describe in general terms. To earn 3 marks, the response must outline a benefit and a limitation of the strategies. If a benefit and limitation are only identified (not outlined), then 2 marks are awarded. Any relevant information not included in the question is awarded 1 mark.

Question 22

a Transcription (1 mark)

b The next step in this process is translation, where the ribosome 'reads' the mRNA strand and builds a polypeptide (1 mark). Ribosomes are located outside the nucleus, in the cytoplasm (1 mark), so the mRNA must leave the nucleus by the path shown in the diagram so that it can reach a ribosome and translation can occur.

Question 23

	Modelling of DNA structure	Modelling of DNA replication
Outline one aspect that is accurate	Base pairing rule is shown using different shapes (1 mark)	Shows semiconservative nature of DNA replication (1 mark)
Outline one aspect that is inaccurate	Does not show alternating sugar–phosphate backbone (1 mark)	Does not show enzymes involved in unzipping (DNA helicase) or synthesising (DNA polymerase) (1 mark)

Note that these are examples of possible answers. There are many other correct answers.

Question 24

a The non-treated diabetic woman's starting glucose level is much higher than the non-diabetic woman's (approximately 290 mg dL^{-1} compared with approximately 100 mg dL^{-1}) and it then rises to a much higher peak of about 475 mg dL^{-1}, compared to about 270 mg dL^{-1} after 20 minutes. The blood sugar of both the non-diabetic and the non-treated diabetic decrease at approximately the same rate, returning to the starting level by 120 minutes. The blood sugar of the insulin-treated diabetic is very similar to that of the non-diabetic woman, with a starting level of approximately 90 mg dL^{-1}, peaking at 280 mg dL^{-1} before returning to the starting level after 120 minutes. While there is a small variation between the insulin-treated diabetic and the non-diabetic, the levels are within the normal range, indicating that the insulin treatment is effective. (3 marks)

For full marks, the response must include:

- a comparison of the non-treated diabetic and the non-diabetic
- a comparison of the insulin-treated diabetic and the non-diabetic
- reference to specific data from the stimulus.

b (3 marks)

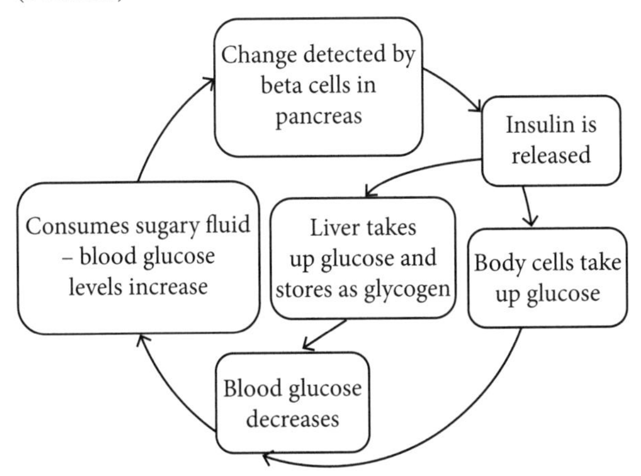

For full marks, The response must include the following points:

- The beta cells of the pancreas detect the stimulus (increase in blood glucose).
- Insulin is released in response to increase in blood glucose.
- Liver stores excess glucose as glycogen and body cells take up glucose.
- Blood glucose decreases.

For 2 marks, the response must include the concept of detection by the pancreas, and release of insulin to lower blood glucose. It may not include details such as beta cells or glycogen. If these two concepts are not included, but some relevant information is present (that is not provided in the stimulus), the response will earn 1 mark.

c Successful educational campaigns provide information and educate the population about the effects of a disease and potential risk factors (1 mark). A program for gestational diabetes should explain what gestational diabetes is and the potential short-term and long-term effects on the mother and child (1 mark). It should also include risk factors for the disease (e.g. maternal age, obesity and other health conditions) and suggestions for how to minimise the risks (e.g. exercise and dietary requirements to maintain a healthy weight during pregnancy) (1 mark).

Question 25

a TT (1 mark)

b The shell thickness appears to follow an incomplete dominance pattern (1 mark), with thick shell not completely masking no shell. This means that the *tenera* parents that the farmer used would both be heterozygous, TN (1 mark). As shown in the Punnett square, two *tenera* parents would produce *dura*, *tenera* and *pisifera* offspring, in a 1 : 2 : 1 ratio (1 mark). It is likely that only 50% of the offspring would be of the *tenera* form.

	T	N
T	TT	TN
N	TN	NN

(1 mark)

c **i** Tissue culture is a form of cloning (1 mark). Because the cells obtained from the root of the parent plant would have the genotype TN, all the offspring will be genetically identical (TN) and *tenera* in form (1 mark).

ii Palm oil agriculture would have an extremely negative impact on biodiversity (1 mark) at all three levels, as it reduces the amount of variation present at the genetic, species and ecosystem level.

For example, genetic diversity refers to the variety of alleles within a species (1 mark). Because oil palm is such an important crop, tissue culture has been used to ensure all palms are of the more productive *tenera* form. The use of tissue culture creates clones that are genetically identical. In this way, agriculture has a negative impact on the diversity of the *E. guineensis* species, as certain alleles that are not desirable are removed from the population as the monoculture is created and maintained (1 mark).

Species diversity refers to the variety of different types of organisms, or species, living within a habitat (1 mark). Palm oil plantations drive the destruction of tropical rainforest, which would normally be home to a vast range of species (rainforests are biodiversity 'hot spots'). Through destruction of habitat, some of the species that are endemic to tropical rainforest ecosystems (e.g. orangutans in Borneo) may become extinct. In this way, palm oil agriculture also has a negative impact on biodiversity at the species level (1 mark).

The verb in this question is 'assess'. 1 mark is allocated to making a judgement about the impact of palm oil agriculture on biodiversity. The response must then provide reasons for this judgement, relating to any two of the three levels of biodiversity (genetic, species and ecosystem). For each level, 1 mark is awarded for demonstrating an understanding of what that level of biodiversity refers to, and 1 mark is allocated for relating this to the palm oil scenario provided. The sample answer addresses genetic and species biodiversity; however, a response could also explain the reduction in ecosystem diversity. This is because different types of ecosystems are lost from a given area (or even on Earth as a whole) because of the establishment of large palm oil plantations.

Question 26 ©NESA 2021 MARKING GUIDELINES SII Q29

The posture of the koala in the tree is associated with the category of hot or mild air temperatures. Posture is a behavioural adaptation that changes the surface area of the skin exposed to the air. In mild conditions, koalas are sometimes observed to curl up, minimising the surface area exposed and therefore minimising heat loss, a behaviour that is not observed in hot conditions. Maximising the surface area of ventral skin exposed by leaning back on a branch occurs much more frequently in hot conditions than in mild.

Hugging of tree trunks is observed more often in hot conditions. Because tree trunk temperature has been shown to be lower than air temperature in hot conditions, this is also a behavioural adaptation by the koala to expose its ventral surface to the cool trunk and use it as a heat sink. (4 marks)

For full marks, the response must:

- demonstrate a through understanding of adaptations to maintain stable body temperature
- explain the relevant adaptations in the koala
- relate the answer to the stimulus provided.

Although not included in the sample answer, answers could include reference to the physiological adaptation of blood flow being directed to the koala's belly to make contact with the tree trunk.

Question 27

a DNA fingerprinting compares the size of DNA fragments from one person to the next (1 mark). Restriction enzymes are used to cut DNA at a specific sequence of bases, called a recognition site (1 mark). Because different restriction enzymes have different recognition sites, using more restriction enzymes will produce a greater number of fragments (1 mark). This increases the likelihood that two different individuals will have unique DNA fingerprints, and therefore increases the accuracy of any conclusions drawn from the fingerprints (1 mark).

1 mark is given for defining a restriction enzyme (i.e. it cuts DNA at a specific sequence of bases). The remaining 3 marks are for explaining why different restrictions enzymes are used. For full marks, the response must link greater number of different restriction enzymes to a greater difference in the fingerprints produced from one individual to the next.

b

1. DNA samples from all five individuals are cut with the same suite of restriction enzymes.
2. The samples are loaded into the wells of the agarose gel at the negative terminal. (1 mark)
3. The power source is used to pass a current through the buffer solution. The DNA (because of its negative charge) migrates towards the positive terminal. (1 mark)
4. The smaller fragments move faster than the larger ones, so they end up closer to the positive terminal. (1 mark)
5. Because of the different sizes of DNA fragments in each individual, a unique fingerprint is produced for all five individuals.

For full marks, a response must include steps 2, 3 and 4 from the sample answer (or words to that effect). Steps 1 and 5 are not required, as they may have already been discussed in part a.

c DNA fingerprinting can be used in forensic investigations (1 mark). DNA samples (e.g. blood) from the crime scene, a victim, and multiple suspects can be compared to determine who committed a crime (1 mark).

Paternity testing is not acceptable as an answer because it is the example provided in the diagram.

Question 28 ©NESA 2021 MARKING GUIDELINES SII Q26

A reduction in gene flow occurs as populations become smaller and more isolated. This results in inbreeding, a reduction in the gene pool and the accumulation of mutations. Many mutations are recessive, and when offspring receive two copies of a recessive mutation it is expressed (and offspring will appear different to their parents). Inbreeding with close relatives results in increased risk of the phenotype being affected by mutations. (4 marks)

For full marks, the response must:

- give reasons for the change in the phenotypes of offspring
- refer to recessive alleles and gene pools in populations
- relate the concentration of recessive alleles to the offspring being different to the parents.

If the response provides reasons and shows a sound understanding of the gene pool, but does not link the different coat patterns to the increased likelihood of recessive alleles combining in offspring, it will still earn 3 marks.

Question 29

a A trait that is controlled by a gene (or genes) on the non-sex chromosomes (chromosome pairs 1–22 in humans) (1 mark)

b **i** The PKD in this family is dominant (1 mark). If it were recessive, II-3 and II-4 would have to be homozygous PKD (1 mark) and if this were the case, they could not have unaffected offspring III-4. This is because the offspring would need to have inherited a dominant allele in order to be unaffected, which neither parent would have (1 mark). Therefore, it must be dominant.

ii As stated in the article, about 35% of people waiting for a kidney transplant die while waiting, or are no longer eligible. This means that people with kidney failure (such as people with PKD) miss their opportunity for a transplant, and dying prematurely due to a shortage of organs.

The use of genetically modified pig kidneys for transplantation could have a hugely positive impact on society, because a person with PKD could come off dialysis sooner, allowing the individual to live with more freedom and be more productive in society. In diseases like ADPKD, where onset of kidney disease is slow and progressive, a suitably matched pig could be on 'standby' until the individual's own kidney function is no longer sufficient. This would bypass the need for dialysis altogether and would reduce the strain on the healthcare services that administer dialysis and the cost associated with it.

A potentially negative social implication exists in terms of access to this potentially life-saving treatment. The cost of the treatment may be a barrier for some individuals, particularly if the treatment is not subsidised by government. This could contribute to a class divide between those who can afford treatment and those who cannot. The treatment may be subsidised in certain cases (e.g. for inherited conditions like PKD) but not others (e.g. for preventable causes such as high blood pressure), and this could have negative impacts on those who are not approved for a transplant.

One potential negative ethical implication is that, as stated in the extract, recipients still require immunosuppressive medication to dampen their immune response and prevent rejection of the pig organ. This leaves the patient susceptible to infection from pathogens that may be present in the pig tissue. This has been shown to occur in monkeys, and could put the life of the patient at risk. Other lower-risk solutions, such as at-home dialysis, or providing incentives for people to become living donors, could be more ethically acceptable from a risk/reward perspective. (8 marks)

For full marks:

- The response must include at least three implications. The question says 'the ethical *and* societal implications', so at least one must be related to society and one to ethics. Each implication must be explained in detail, using information from the stimulus as evidence to support any implications drawn.
- The verb is 'discuss'. Because there are clear positive and negative implications in this case, at least one implication must be positive and one negative.
- The response must be coherent, with a logical progression of ideas.

There are many potential implications that could be discussed. Others answers may include the following points:

- Some cultures and religions reject the idea of using pig tissue, which may lead to exclusion of some groups and unequal access in society to this potentially life-saving technology.
- Determining which patients are eligible to access to xenotransplantation and when the decision is made for each patient could pose ethical concerns for doctors and patients.
- The risk of zoonotic infection extends to broader society, as a zoonotic pathogen may spread throughout the human population. This presents both societal and ethical implications in terms of whether it is fair to risk the health of many in order to improve the health of a few.

Question 30

a Base substitution / point mutation (1 mark)

b Mutations that occur in somatic cells only affect the individual (1 mark) in whom the mutation took place. Mutations must occur in the gametes (sperm or ova) in order for the offspring to be affected (1 mark). The cervical cells are somatic cells, and therefore the mutation would not be passed on to the offspring (1 mark).

c An infectious disease is one that is caused by a pathogen and is transmitted from one host to another (1 mark). Some may consider cervical cancer to be infectious as it occurs in the presence of HPV infection. However, the virus does not directly cause the cancer, but rather acts as a mutagen, causing mutation in the cells, which leads to cancer (1 mark). Thus, cervical cancer is a non-infectious disease (1 mark).

Responses should show an understanding of the definitions of infectious and non-infectious disease, recognise that the virus is a mutagen rather than a pathogen, and therefore conclude that it should be considered non-infectious.

d Vaccination against HPV was introduced in 2007 and has been successful in reducing the incidence of cervical cancer. Extending the vaccination program to include males has reduced the rate of transmission, further reducing the incidence. It is predicted that vaccination alone, without continued screening of the population, would allow a reduction in incidence from about 7 cases per 100 000 women to 2 cases per 100 000 by 2040, where the rate would remain relatively stable. The mortality rate is also predicted to be reduced by vaccination from about 1.75 per 100 000 in 2020 to about 1 by 2090.

In every population there are some people who cannot be vaccinated because of compromised immunity or medical treatments. High rates of vaccination in the community will provide 'herd immunity', protecting these vulnerable people, as so few people will be susceptible to the disease that it is not likely that they will be exposed.

From 1991, when screening programs for cervical cancer were introduced in Australia, the incidence and mortality rates for cervical cancers dropped dramatically, leading to one of the lowest rates in the world. The graph shows a slight increase in incidence after the screening test was changed in 2017 to test for HPV rather than cancerous cells. This is due to a more sensitive test picking up cases earlier and coincides with a decrease in mortality rate as early detection of cancer improves outcomes for treatment and survival.

The graphs show a prediction that the incidence and mortality rates for cervical cancer would be much lower with a screening program than with vaccination alone. Incidence is predicted to continue to decrease to a rate of about 0.5 cases per 100 000 by 2090 (compared to a steady rate of about 2 per 100 000 after 2040 with vaccination alone). This is likely to be due to having fewer undetected cases and thus reduced transmission of the disease.

The mortality rate would decrease even more, to about 0.1 cases per 100 000 with a screening program by 2090, compared with 1 case per 100 000 with vaccination alone. With cases detected earlier, treatment is more successful and deaths will be reduced.

It can be seen from this information that both vaccination and screening programs are highly effective at reducing the incidence and mortality of cervical cancer. However, the best predicted outcomes occur when both vaccination and screening programs are offered. Therefore, it would be most beneficial for both programs to continue. (9 marks)

This question can be approached in different ways. The response should include:

- analysis of the information and data presented for vaccination programs, both past and future
- an understanding of the benefits of herd immunity for the unvaccinated
- analysis of the information and data presented for cervical cancer screening programs, both past and future
- specific data included from both graphs, for vaccination alone and for vaccination combined with a screening program
- an evaluation of the success of vaccination and screening programs
- an evaluation of the need for the programs to continue
- use of scientific terminology.

9780170465250